MOBILE INTERFACES IN PUBLIC SPACES

With today's mobile devices, not only can we connect to the Internet anywhere at any time, we can also map our precise geographic coordinates and access location-specific information that enables us to engage with our surroundings in new ways. The proliferation of location-aware mobile technologies calls for a new understanding of how we define public spaces, how we deal with locational privacy, and how networks of power are developed today.

In *Mobile Interfaces in Public Spaces*, Adriana de Souza e Silva and Jordan Frith examine these spatial and social changes by framing the development of location-aware devices within the context of other mobile and portable technologies. These technologies work as interfaces to public spaces—that is, as symbolic systems that not only filter information but also reshape communication relationships and the environments in which social interaction takes place. de Souza e Silva and Frith suggest that far from alienating us from our surroundings, mobile interfaces can foster deeper connections with the people and places around us.

Adriana de Souza e Silva is Associate Professor in the Department of Communication at North Carolina State University. Her research focuses on how mobile and locative interfaces shape interactions with public spaces and create new forms of sociability. She is the co-editor of *Digital Cityscapes: Merging Digital and Urban Playspaces*, and co-author of *Net-Locality: Why Location Matters in a Networked World*.

Jordan Frith is a doctoral candidate in North Carolina State University's Communication, Rhetoric and Digital Media program. His main research interests are locative media and space, particularly how locative media may influence interactions in urban spaces. He has recently been published in the journals *Mobilities* and *Communication, Culture, and Critique*.

MOBILE INTERFACES IN PUBLIC SPACES

Locational Privacy, Control, and Urban Sociability

Adriana de Souza e Silva
and
Jordan Frith

Routledge
Taylor & Francis Group

NEW YORK AND LONDON

First published 2012
by Routledge
711 Third Avenue, New York, NY 10017

Simultaneously published in the UK
by Routledge
2 Park Square, Milton Park, Abingdon, Oxon OX14 4RN

Routledge is an imprint of the Taylor & Francis Group, an informa business

© 2012 Taylor & Francis

The right of Adriana de Souza e Silva and Jordan Frith to be identified
as authors of this work has been asserted by them in accordance with
sections 77 and 78 of the Copyright, Designs and Patents Act 1988.

Library of Congress Cataloging in Publication Data
Silva, Adriana de Souza e.
  Mobile interfaces in public spaces: locational privacy, control, and urban
  sociability/by Adriana de Souza e Silva & Jordan Frith.
    p. cm.
  Includes bibliographical references and index.
  1. Mobile computing.   2. Location-based services—Social aspects.
  3. Social media.   4. Privacy, Right of.   I. Frith, Jordan.   II. Title.
  QA76.59.S55 2012
  004—dc23   2011038996

ISBN: 978-0-415-88823-3 (hbk)
ISBN: 978-0-415-50600-7 (pbk)
ISBN: 978-0-203-12396-6 (ebk)

Typeset in Bembo and Stone Sans
by Florence Production Ltd, Stoodleigh, Devon

To my mom, Elizabeth, with love.
Adriana

To my family and friends. I wouldn't be here without you.
Jordan

CONTENTS

FIGURES

ACKNOWLEDGMENTS

The acknowledgements page is generally one of the hardest parts of a book to write. It's not that it is hard to thank others, but we always fear that we will forget someone who has offered help and has contributed to the realization of this project. We are very glad to have had so much support during the process of writing this book. First of all, we would like to thank our home institutions, North Carolina State University and the IT University of Copenhagen.

We both would like to thank Amani Naseem for her laborious help in getting permission for the images in this book. She has always been very reliable and competent, and we are immensely grateful for that.

A special thanks for Erica Wetter and Felisa Salvago-Keyes at Routledge, who worked with us since the beginning of this project, patiently answering our many questions, and promptly supporting its development so that this book could reach its full potential. Also thanks to our reviewers, who greatly helped us improve the manuscript.

We also would like to acknowledge the feedback we received during conferences where we presented chapters of this book along the way. Particularly, the National Communication Association conference in San Francisco (2010), the Homo Ludens 2.0 conference in Utrecht (2010), the Mobilities in Motion conference in Philadelphia (2011), and the Transforming Audiences 3 conference in London (2011).

Adriana

I particularly would like to thank the IT University of Copenhagen for having provided me with the perfect research environment when we were fiercely writing these chapters. Here I refer not only to the time provided for writing, but also the contact and conversations with scholars who inspired me and challenged some of my assumptions about technology. Particularly, I would like to thank Rich Ling, Irina Skhlovski, Gitte Stald, Lone Malmborg and John Paulin Hansen.

Jordan, thanks for working with me on this project, you have always been so much on top of our deadlines and provided me with valuable insights and great feedback. Thanks for making this one of the easiest and most pleasant scholarly collaborations I've had.

Finally, I would like to thank my husband John for the ongoing support and immense love. He was never tired of engaging in conversations with me about location-based technologies and of repeatedly reading my text whenever I asked for it.

Jordan

I want to thank my colleagues at NC State, especially my cohort. I chose to come here because of all of you, and I've never regretted it.

I want to thank Adriana for being as good a mentor as any doctoral student could possibly wish for. I wouldn't have had this opportunity if it wasn't for you, and you've been amazingly understanding throughout the entire process. At this stage in my career, I know I've probably been spoiled by how easy it has been to work with you.

Also I want to thank my mother for putting up with my refusal to ever talk about my research. Hopefully, you'll read this book and find out what I've been up to for the last year.

Adriana and Jordan
September 2011

INTRODUCTION

In his book *Speaking into the Air*, John Durham Peters (1999) traces 2,000 years of thinking about communication, ranging from Socrates to Alan Turing. The story he tells is of philosophers forced to face one of the most glaring imperfections of the human condition: People cannot directly communicate with one another. Even if they grow up together, know each other's history, and could trace each other's life trajectory, the thoughts in one's head cannot transfer directly to somebody else's. The point seems obvious, but the imagining of a perfect form of communication is alive and well. We can perceive it whenever someone contrasts face-to-face communication with "mediated" communication. But what exactly is "mediated communication" anyway? Is there such a thing as unmediated communication? Even face-to-face communication is mediated by bodies, language, and a whole set of neural processes. As Peters points out, the reality of unmediated communication would be quite unpleasant. Philosophers may bemoan the impossibility of pure communication, but without the filters that make direct communication impossible, people would not be able to interact with each other.

We have always needed different types of filters to interact with the world. Our bodies, language, signs, and symbols are just some examples of these filters. There have never been unmediated interactions with others or with the spaces that surround us. But the relationship between us, the world around us, and what makes this relationship possible have been far from constant. As social norms and social spaces change, we have developed new types of filters and devices to interact with others and our environment. These filters and devices, and their affordances and constraints, have changed throughout history.

In the 1960s, a new term to describe some types of filters was popularized: *interface*. Broadly speaking, *interface* means something that is between two other parts or systems, and helps them communicate or interact with each other. Interface is something that makes a connection between two parties, but it also becomes part of that system, influencing how they interact with each other. According to the Oxford English Dictionary, although the term existed since the 1880s, "in the 1960s it became widespread in computer use and, by analogy, began to enjoy a vogue as both a noun and a verb in many other spheres" (Simpson, 1999). It was not until the 1980s, however, that the term left the computer science world and became a part of the everyday life of common users. The Macintosh's (and later Microsoft Windows') graphical user interface (GUI) brought the term into the popular domain. At its simplest level, the GUI's primary task is to re-represent (or translate) the computer's binary code language of 0s and 1s into a visual language that is understandable to non-specialized users. Computers have always been machines capable of representing information. Early computer programs translated basic computer binary code into a programming language that was understandable to programmers and vice versa. But the GUI added another layer of representation to this dynamic, re-presenting the programming languages into icons and images, and transforming the computer screen into a space to be navigated and inhabited.

The idea of computers as space is older than the GUI, however. In the 1960s, Douglas Engelbart invented the mouse and was the first to envision the computer screen as a space to be navigated (Johnson, 1997; Lévy, 2004; Rheingold, 1993). The mouse pointer was the first human representation on the computer screen, and as such, it was also the first avatar—one of the earliest computer interfaces. Pierry Lévy (2004) describes the development of the personal computer as the process of adding layers of interface to each other, such as the mouse, the keyboard, the screen, the operating system, the hard drive, and eventually the GUI. Lévy's narrative intended to show that computer interfaces do not only translate or mediate—rather, they play a critical role in shaping interactions and creating meaning. Each additional layer not only contributed to shape how we understand computers today, but also fundamentally changed how people interacted with computers—from a calculating machine to a space to be navigated and inhabited. When Sherry Turkle (1995) described the transition from the culture of calculations to the culture of simulations,[1] she showed that the new interface layer of the GUI not only altered the meaning of computers, but also the minds of the users who interacted with computers. In Turkle's words, "We construct our technologies, and our technologies construct us and our times. Our times make us, we make our machines, our machines make our times. We become the objects we look upon, but they become what we make of them" (1995, p. 46).

Computer interfaces are not neutral. They actively influence communication relationships (in this case, the relationship between a human and a computer), and transform both parties that it connects. Generally speaking, computer interfaces are software and hardware that allow communication between computer systems and human users, a term labeled "human-computer interface." In the case of the human-computer interface, the role of the interface is to translate digital information from computers to humans in order to make it understandable to us. Nicholas Gane and David Beer (2008), however, show that the primary technologies associated with "interface" shifted in the last few years from computer interfaces (most notably represented by the desktop GUI) to what they call pervasive interfaces. Pervasive interfaces are mobile and "invisible" devices, such as mobile phones, RFID tags, and other location-based technologies. These interfaces that enable people to track others and to be tracked have, according to Gane and Beer, important implications for our sense of privacy, and influence surveillance, control, and power mechanisms in today's society. And because these mobile interfaces are becoming increasingly ubiquitous, interfaces should be understood as spatial forms that are tied to broader sets of social and cultural dynamics (Gane & Beer, 2008)

Steven Johnson (1997), indeed, proposes a definition of interface that goes beyond digital technologies. For Johnson, what characterizes an interface is not its digital characteristics, but rather the fact that it works as a translator, a mediator between two instances, making one part sensitive to the other. Lévy (2001) supports a similar definition, suggesting that the term "interface" should not be limited to digital technologies. He identifies the printed book in the fifteenth century as an important type of interface that changed how knowledge and information were transmitted, received, and organized. If we understand interfaces broadly as producers of meaning, and ways of representing and organizing knowledge, information, and space, then older technologies, such as books and film cameras are also interfaces because their role is to represent the author's words to readers (in the first case) or to translate the external world into moving image frames (in the second case).[2] Lev Manovich (2002) also supports this argument, identifying not only the GUI, but also books and cinema as cultural interfaces. The notion of cultural interface means that material devices are not only mediators between a person and a machine, but, most importantly, they are also filters for culture, defining and influencing how people interact with the world around them.

Following this logic, Gane and Beer (2008) point out three major ways of conceptualizing interfaces today: they are cultural devices; they mediate everyday experiences in social and physical spaces; and they enable different forms of power and/or surveillance. This tri-faceted view of the notion of interface is important because it allows us to think beyond the traditional dualism man vs. machine, and it understands interfaces as systems that include the user, as well as the space around the user. For Gane and Beer, "Interface is the meeting point of a number of important social and cultural dynamics, for it enables and mediates informal power structures, restructures everyday practices in a myriad of ways, and transforms relations between bodies and their environments" (2008, p. 61). They further suggest that

> Perhaps the best way forward is to avoid thinking of interfaces in isolation as discrete technologies in their own right, and to conceive of them instead as membranes or mediating devices that are not only integrated into our everyday routines, but tied to a deeper set of social and cultural processes.
>
> (2008, p. 61)

So, more than simply translators or communication mediators, more than specific technologies, interfaces are symbolic systems that filter information and actively reshape communication relationships, and also reshape the space in which social interaction takes place. As such, interfaces influence our perceptions of the space we inhabit, as well as the types of interaction we have with other people we connect with in these spaces (de Souza e Silva, 2006). This framework leads to the concept of "social interface" (de Souza e Silva, 2006, p. 261), which focuses how interfaces are connected to their place of operation and how they frame everyday sociality.

This book is about interfaces.

But it is not about the typical computer interface, or the well-known GUI. It is about the types of interfaces that individuals carry around in public spaces, which we call personal mobile technologies. Applying the idea of social interface to personal mobile technologies allows us to think of these devices as more than simple technologies, or material devices, and rather as filters, control devices, organizers of social networks, locative technologies, and information access platforms. Not all interfaces discussed in this book will perform all of these functions at once. We analyze some primarily as filters, and we discuss others mostly as control devices and information access platforms. More often, however, the same interface performs different roles depending on how it is used. For example, an iPod can be used as an information (music) access device, but also as a way to filter social interactions in public spaces. Mobile phones additionally might enable the organization of social networks. However, what is important to consider is that "every shift in the meaning of an interface requires a reconceptualization of the type of social relationships and spaces it mediates" (de Souza e Silva, 2006, p. 262). This is crucial to our broader understanding of interface as a system that includes material devices, people, and spaces.

In this book, we specifically address five kinds of mobile interfaces: the book, the Walkman, the iPod, the mobile phone, and location-aware technologies.[3] All these technologies have the common characteristics of being easily portable, generally designed for individual use, and often carried outdoors and in public spaces. Paraphrasing Mizuko Ito, Daisuke Okabe, and Misa Matsuda (2005), they are all personal, portable, and pedestrian. As interfaces, they significantly influence how people perceive and interact with public spaces and with other people in these spaces. Although the adoption and use of these technologies is intrinsically connected to the spaces in which they are used, they have often been blamed for disconnecting users from their surrounding spaces and from the people around them (du Gay, Hall, Janes, Mackay, & Negus, 1997; Gergen, 2002; Geser, 2004; Puro, 2002; Turkle, 2011).

The argument goes something like this: "The Walkman/iPod/mobile phone makes people less likely to talk to strangers and pay attention to their surrounding environment." It would not be accurate to argue that these technologies do not have at least some effect on urban sociability, but we do not need to fall back on technological determinism to do so. Manuel Castells (2000) emphatically avoids the technological determinist perspective, emphasizing that the "dilemma of technological determinism is probably a false problem, since technology *is* society, and society cannot be understood or represented without its technological tools" (p. 5). The idea that technology and society should not be viewed as separate and opposite to each other is indeed critical for the understanding of the interrelationships between mobile technologies, people, and public spaces as an interface system, and thus relevant for our approach in this book. If we take this perspective, it is difficult to attribute complete responsibility to mobile technologies for bringing us together or pushing us apart. Instead, a more productive way to analyze these technologies is to understand how they work as interfaces in the urban environment that actively interact with users to shape communication relationships and public spaces. Ultimately, this book is about how mobile technologies can be viewed as *interfaces to public spaces*, that is, systems that enable people to filter, control, and manage their relationships with the spaces and people around them.

Since the popularization of the paperback novel, people have carried mobile technologies in public spaces, such as plazas, cafes, and trains, and used these technologies, among other things, to pay selective attention to their environment or to avoid interacting with strangers. For example, Wolfgang Schivelbusch (1986) shows that the paperback novel became popular around the same time that people started traveling by train. By reading a book, passengers did not feel forced to interact with strangers sharing the same railway compartment. However, as we will see in the first two chapters of this book, the desire to avoid interactions with strangers and the need to selectively filter the over-stimulation of the urban environment is not only a consequence of mobile technology use, but also the result of the growth of the modern city at the beginning of the nineteenth century, among other factors. At the turn of the century, Georg Simmel (1950) noted that urbanites developed a *blasé* attitude in order to pay selective attention to the urban environment that was increasingly full of external stimuli. People walked around the city, seeming at the same time indifferent and anonymous to the crowd around them, so they could control how they engaged and disengaged with their environment. We still do that today. When walking on a crowded street in a big urban center, we consciously (or sometimes unconsciously) pay attention to some things and not others. We give a coin to a homeless person begging for money, but we ignore a street performer. Or we pretend we do not see that acquaintance crossing the street corner because we are late for a meeting. Loosely speaking, the *blasé* attitude can be considered an early type of interface—a psychological filter and a way of managing and controlling interactions with the city.

Often discussions of mobile interfaces, particularly mobile phones and digital music devices, forget to take into consideration a broader framework that situates these technologies into the spatial and social development of contemporary urban life. If we want to pretend we do not see an acquaintance crossing the street corner, we look down as if we are late for a meeting—or we turn on our iPods to signal we are less approachable. People talking with a friend or a family member on their mobile phones on a bus or a street bench have been criticized for missing the opportunity to talk to the stranger sitting besides them (Turkle, 2011). But would they talk to them had they not had the phone? Can we really say that this tendency is purely a consequence of technology use? One of the goals of this book is to show that desires of managing and controlling our interactions to people around us and to the places we inhabit are older than the popularization of the Walkman, mobile phones, and iPods. More importantly, people do not detach themselves from the places they physically inhabit, but rather develop new types of relationships with them. If place is where we inscribe personal meaning, as Edward Casey (2009) argues, then we are still "in place" when we walk down the street listening to head phones or talking on a mobile phone. It becomes, however, a different place with different inscribed meanings.

The different meanings inscribed to a place will be even more evident with the popularization of location-aware mobile technologies. Location-aware technologies are mobile devices able to locate themselves via global positioning system (GPS), Wi-Fi, or triangulation of radio waves, and therefore able to provide users with location-specific information. Although mobile phones have always been somewhat location-aware, they increasingly come with GPS capabilities and operating systems that can run applications (apps) that allow users to locate other people or find the location of things and information around them. These types of phones are popularly known as "smartphones." The ecology of location-aware technologies also includes Bluetooth and GPS devices able to represent and transmit location-based information. Ultimately these technologies enable users to attach information to places by means of latitude/longitude (lat/long) coordinates.

While other types of mobile technologies were often blamed for making their users not pay attention to their surrounding environment, either by looking at a printed page, by listening to sounds that did not belong to the city, or by talking to somebody remote (Bull, 2000; du Gay et al., 1997; Katz & Aahkus, 2002; Moores, 2004; Plant, 2001), location-aware technologies strengthen people's connections to their surroundings because they help people to locate other people and things around them (de Souza e Silva, 2006, 2009; de Souza e Silva & Sutko,

2009; Gordon, 2008; Humphreys, 2007). For example, location-based social networking (LBSN) software such as *Foursquare* and *Gowalla* allow users to see the location of their friends on their mobile phone screen. Likewise, location-based advertising (LBA) can deliver coupons whenever a user is within a certain radius of a specific store, and location-based services (LBS) such as *WikiMe* and *GeoGraffiti* allow individuals to access and upload information that is place-specific. Differently from other mobile technologies, location-aware technologies are types of computer interfaces in that the mobile phone screen graphically represents digital information to its users—they are mini-computers. But they are also very particular types of interfaces, because they let users visualize their surrounding space through a map displayed on the screen. Location-aware technologies however, do not represent all the surrounding space. They take aspects of that space (certain types of information, certain types of users) and selectively display them on the cell phone screen depending on users' preferences. Consequently, location-aware interfaces promote important new ways people can filter public spaces.

This book is about mobile technologies as interfaces to public spaces.

With the increasing pervasiveness of location-aware technologies, public spaces are altered by the proliferation of location-specific digital information. When somebody carries a book, a Walkman, or an iPod, their perception of space is changed, but the space itself remains unchanged for other people in the surroundings. A person walking around the neighborhood of Beverly Hills in Los Angeles who listens to the soundtrack of *Beverly Hills Cop* on his iPod might imagine himself in a scene from the movie. But unless the iPod user starts to sing aloud, everybody else on the streets is oblivious to that imaginary scene. Location-aware technologies, on the other hand, allow people to attach information to locations, changing that space for other people using those services. For example, many LBS allow users to upload location-based pictures, attach reviews to restaurants, or post tips about public spaces that can be then accessed by their users.

More importantly, the popularization of location-aware technologies contributes to the changing meaning of locations. With the increasing use of location-based services, location acquires relevancy as it fundamentally shapes social, political, and spatial interactions: People check into locations, they are concerned about locational privacy, they have their location tracked, they attach information to locations, and they are even able to "create" new locations. For example, *Foursquare* users can add new locations to the game, to which other users can "check in."[4] A player can transform his bathroom into a new location, and another player can become the mayor of his backyard—if she defines it as a new *Foursquare* location. But locations are not only trivial spots. A Dallas citizen for instance, might define the spot President Kennedy was shot as a location, and other visitors to Dallas will be able to contribute to the informational landscape of that location with comments, testimonies and pictures. Locations become filled with a wealth of digital information that can be accessed, used, and added to by others. With location-aware technologies, *location* becomes paramount to social interaction and spatial construction, and a critical concept that frames our analysis of these types of mobile technologies.

The concept of location has generally been defined in relation to the idea of place. In geographical terms,

> Location is the position of something, expressed in grid-coordinates, in relation to other things, or in such terms as near and far. Place denotes a site or portion of space in which objects and personas are often intimately related, as a setting for individual activity and social interaction.
>
> (Smith, 2000, p. 45)

In the quote above, David Marshall Smith emphasizes the social characteristics of places, an aspect that has also been highlighted by many space/place scholars (Harvey, 1993; Lefebvre, 1991; Massey, 1995). Conversely, locations are merely defined by the existence of lat/long coordinates. As Tim Cresswell (2004) notes, locations have "fixed objective co-ordinates on the Earth's surface" (p. 7). Although places can also be located, as Cresswell shows, they are not always stationary. He employs the example of a ship to point out a type of mobile place for people who are on a journey. A ship is a place, "even though its location is constantly changing" (Cresswell, 2004, p. 7). A ship on a journey then, is a place that moves through many different locations.

The notion of location, however, has also often been subordinated to the idea of place, conceptualized as an aspect of a place, deprived of meaning (Agnew, 1987; Cresswell, 2004; Harvey, 1996; Tuan, 1977). As Ginette Verstraete and Tim Cresswell (2002) note, "location refers to an abstract point in abstract space. As such it is devoid of meaning and cultural significance" (p. 12). David Harvey (1996) corroborates this idea by affirming that places have "a discursive/symbolic meaning well beyond that of mere location, so that events that occur there have a particular significance" (p. 293). According to this logic, places are meaningful and filled with social relationships. In contrast, locations have only a geographical position. This notion of place as socially meaningful has often been established not only in contrast to location, but also to space. Following Michel De Certeau (1988), Steve Harrison and Paul Dourish (1996) defined a place as "a space which is invested with understandings of behavioral appropriateness, cultural expectations, and so forth . . . 'Places' are spaces that are valued" (p. 69). And the geographer Yi-Fu Tuan (1977) suggested, "When space feels thoroughly familiar to us, it has become place" (p. 73).[5] Yi-Fu Tuan was one of the first scholars to emphasize the relevance and meaning of place, which led to renewed scholarly interest in that concept. Places thereby acquired new significance among scholars, as they were viewed as meaningful spaces, constructed by social forces (Agnew & Duncan, 1989; Massey, 1993; Sack, 1992).

We suggest that we are now witnessing, perhaps with the same or greater intensity, a renewed interest in the idea of location. The popularization of location-aware mobile technologies not only highlights the importance of location, but also forces us to re-think how location has been traditionally conceptualized. Locations are still defined by fixed geographical coordinates, but they now acquire dynamic meaning as a consequence of the constantly changing location-based information that is attached to them. For example, visitors to the city of Copenhagen who open the location-based game *The Gold Horns Thief* on their mobile phone will see "tracks" that tell them about the theft of the Golden Horns, a piece of Viking art that was stolen and melted in the early nineteenth century.[6] Visitors can choose to follow the tracks by solving riddles, and thus advance as the main character in a fictitious story that involves finding where the thief is hidden in the city. Each location that players visit triggers different riddles and bonus files that contain videos about some peculiar aspect of the history of Copenhagen in the specific location where they stand. Players can also access information left by previous visitors, such as pictures and comments, which become permanently attached to those locations. The meaning of each location in the game is thus constantly shifting and being constructed by the increasing amount of location-based information that is attached via location-aware mobile technologies. As a result, finding a location no longer means only finding its geographic coordinates, but also accessing an abundance of digital information that now belongs to that location. This information is always dynamic because it is constantly being created, deleted, and edited. Additionally, the information that belongs now to locations is user-specific. Depending on the type of hardware and software a person has, she will access different aspects of that location.

The dynamic nature of locations can be understood as an interplay between the actual and the virtual.[7] Adriana de Souza e Silva and Daniel M. Sutko (2011) suggest that location-based applications are interfaces that allow each user to actualize some (but not all) aspects of a location. For example, when using an LBS such as *AroundMe*, a person can find nearby restaurants. But if somebody else opens the same application in the same location, but searches instead for coffee, she will get different results on her mobile phone screen. Both people are using the same application, but because they establish different search filters, they will download different information from the database of information that is around them—but never the whole database. *AroundMe* recommendations are already linked to that location, but they remain virtual until accessed via the *AroundMe* software by a user on a mobile phone. As de Souza e Silva and Sutko (2011) suggest, some of these recommendations remain virtual, because a user does not have access to them, either because she does not have the technology or because she used different search criteria. From the perspective of the user there is a dynamic articulation between digital information and location, an articulation that becomes actual through the use of LBS and location-aware technologies. Take the example of two people in the same location, one looking for budget meals and another looking for expensive restaurants using an LBS such as *Yelp*. Even though both are in the same location, they access different information through the location-based app. As such, location-aware technologies interface between location and digital information so that changes in one are also changes in the other (de Souza e Silva & Sutko, 2011). In the previous *Yelp* example, if a person reads a bad restaurant review, this could literally cause a change in physical location for her, in case she decides to go to another street or neighborhood to find a better place to eat. Similarly, a change in the user's physical location leads to accessing different information because location-based information is only displayed within a specific radius of the user's location. This new notion of location allows us to think of location, the user, and the location-aware technology as a unit, or in other words, as a set of interfaces, because they mutually influence each other. Therefore, a change in one will cause changes in the other two.

This interplay between digital information, location, people, and location-based information highlights the fact that locations do influence the way we access and interact with (digital) information. Now geographical coordinates are attached to digital information (e.g., texts, pictures, videos, and people's locations) that is produced and consumed by people in that location. As a result, locations become dynamic and embedded with digital information, acquiring discursive and symbolic meaning, and cultural significance.

Locations, however, are different from places. As we have previously seen, not all places have fixed geographical coordinates. For example, heaven and hell are places, but cannot be located. Moreover, as we will see in Chapter 6, locations now have identities of their own; they are not merely subsets of places. During the past decade, the term location has steadily acquired increased significance in specifically describing the locus of social and spatial interactions that emerge from the use of location-aware technologies. Locations, previously places deprived of meaning, take on complex, multifaceted identities that expand and shift according to the information inscribed in them. It is not that locations become places (after all, places do not necessarily have location-based information embedded in them), but instead *many places do become locations*, that is, the locational aspects of many places acquire relevance. For example, the National Museum, in Copenhagen has long been a well-known place. But recently the museum staff have created a Wikipedia page about the history of the museum that can be accessed via *WikiMe* when visiting the museum. The museum cafe also has reviews that visitors can access when in the museum's vicinity. Finally, the *Gold Horns* game described above ends at the museum, and *Gold Horns* players can learn about the history of the museum via this location-based game. All this information was given lat/long coordinates and inscribed in the space of the museum, thereby transforming it primarily into a location. *The museum is still a place, but its locational aspects become increasingly more significant in shaping the ways visitors interact with it.* As the amount of location-based information grows, many visitors will refer to the museum as a location.

The previously mentioned applications also remind us of how the different apps frame the museum as a location in varying, frequently overlapping ways. *Foursquare* and *Gowalla* users may check in at the museum. However, other visitors will be able to create other locations within the museum space. A *Foursquare* player might designate the museum bathroom as a location so that he can become its mayor. The room that contains the replica of the Golden Horns becomes another location through the *Gold Horns* game. We can thus see that the museum becomes increasingly saturated with different locations, each with its own identity and character.

What we want to emphasize here is that when combined with location-based information, locations are gaining relevance and acquiring statuses similar to, yet distinct from, those of space and place as mediators of our social, cultural, and spatial interactions. We also want to reiterate that location-aware technologies and the information they interface with are not *outside* locations, that locations are not disconnected from the location-based information now embedded in them, and that people are not purely mediators between locations and LBS. Instead, they are all a *set of interfaces*, mutually influencing each other. Location, digital information, locative applications, and users are all tightly bound together. The removal of one from the set of relationship changes completely the nature of this relationship. The recognition of this shift in character of locations has major implications for how we navigate urban spaces, socialize with others in the city, and for how we understand privacy, surveillance, and exclusion.

This book is also about locations, and how locations are transformed by the emergence of location-aware technologies.

The use of location-based applications and location-aware technologies contributes to individualize people's perceptions of public spaces. As mentioned earlier, people use different types of applications to access various types of information, which makes their perception of public spaces more filtered and personalized. The public space through which they move will be perceived different for those who do not possess location-based technologies, or those who do have access to the technology but choose not to engage with it. Although differential perceptions of public spaces always existed, it is important to question how the increased use of location-aware interfaces in public spaces will influence the configuration and perception of public spaces, both to users and to non-users of these technologies.

Besides differential perceptions of public spaces, there are also other important social and spatial implications of the use of location-aware technologies that have not been fully addressed in the literature. Much more than other types of mobile technologies, the pervasive use of location-aware devices raises important concerns about privacy and power in public spaces. Although other mobile technologies such as the Walkman and the mobile phone were claimed to help users "privatize" public spaces, they did not directly threaten users' privacy. When talking on the phone or listening to music with headphones, people still could be anonymous in public spaces: They did not fear that their personal and location information could be accessed just because they were carrying a mobile technology. Location-aware technologies, however, enable users to locate things in space, but in turn they can also be located. Being located in itself is not a problem; many people actively choose to participate in LBSNs such as *Foursquare* and *Gowalla* with the goal of telling their friends where they are. Many people are also willing to let a service such as *Where* know their location so they can easily find gas stations and restaurants nearby (Barkhuus & Dey, 2003; Hong et al., 2003). But concerns arise when people do not know who has access to their location information (de Souza e Silva & Frith, 2010). For example, a *Facebook Places* user might think she is sharing her location only with specific friends in her social network. But when an acquaintance contacts her via the location-based social network application and claims to know where she is, or when she receives unrequested location-based advertisements, it becomes clear that her location information was shared with unauthorized parties. When this happens, people often feel vulnerable, because they do not know how their location information is being disclosed. Although sometimes safeguarding locational privacy is directly related to the ability to understand and adjust the application's privacy settings, often privacy settings are not clear enough to allow users to make an informed decision about the disclosure of their location information.

The disclosure of location information is not only a concern because it threatens locational privacy, but also because it can strongly reinforce or reshape power relationships. For example, parents are increasingly tracking children in order to know their whereabouts and make sure they are "safe." Parents have always monitored their children, and mobile phones have received a lot of attention as technologies that help parents to be in constant touch with their children in order to coordinate daily activities and to know where they were (Devitt & Roker, 2009; Ling, 2004; Ling & Yttri, 2002). But with GPS technology, although monitoring becomes less obvious (no need for constant phone calls), it is nonetheless constant. Also, since 2006 the state of California in the United States uses GPS technology to track sex offenders (Shklovski, Vertesi, Troshynski, & Dourish, 2009). These parolees know parole offices are constantly watching them, an awareness that leads to new ways of disciplining behavior. These two dissimilar examples raise a similar question: Both children and parolees are apparently more mobile, but are they more free? As Doreen Massey (1993) argues, "Mobility and control over mobility both reflect and reinforce power. It is not simply a question of unequal distribution, that some people move more than others, some have more control than others" (p. 62). As such, location-aware mobile technologies are reshaping mobility, privacy, power, and control in public spaces, as well as social relationships.

This book is about privacy and power.

Privacy and power issues have been frequent concerns with the emerging use of location-aware technologies. However, there is still a lack of theoretical and empirical explorations on the social implications of these technologies. To date, there are some studies from a human-computer interaction (HCI) perspective that focus on privacy and power issues related to the use of location-based technologies, but these studies tend to focus on a small number of users and their goal is technology development, rather than understanding social practices and human behavior associated with technology (Barkhuus & Dey, 2003; Boesen, Rode, & Mancini, 2010; Consolvo et al., 2010; Consolvo et al., 2005; Hansen, Alapetite, Andersen, Malmborg, & Thommesen, 2009; Hong et al., 2003). At this early stage of development, critically analyzing the social use of location-aware technologies is crucial for shaping our understanding of how these technologies will be integrated into society and how they will influence social norms and policy-making processes, simultaneously empowering users to control their environments while also being controlled by others.

While the first two chapters do not directly deal with location-aware technologies, the idea of mobile technologies as filters and control mechanisms that help people interact with public spaces and other people in them is present throughout the book. The ability and desire to control interactions with public spaces is not new and did not emerge with location-aware technologies. From the book to the iPod, personal mobile technologies have helped individuals engage with the public on their own terms. For instance, by reading a book on a crowded train, people pay selective attention to their physical surroundings (Schivelbusch, 1986). The same goes for the iPod and the Walkman: By adding an auditory layer to public spaces, users are able to control their otherwise "chaotic" interaction with urban spaces (Bull, 2007; Hosokawa, 1984). These issues of personalization and control over space were present with earlier mobile technologies; however, with location-aware mobile technologies they become more pronounced. As we mentioned above, with location-aware mobile technologies, people can sort and filter information in new ways. Consequently, they have more control over how they interact with physical spaces and others in those spaces, but they are also faced with new power, privacy, and surveillance issues, that is, the threat of being controlled by somebody else. Paradoxically, the increased control over space sits in an uneasy tension with the lack of control over location information.

This book is ultimately about control.

With the increased popularization of location-aware technologies, especially after the release of the GPS-enabled iPhone 3G and Google's Android operating system in 2008, issues of personalization, control, surveillance, and privacy related to location-aware mobile devices have entered the discourse of mass media outlets and have become common topics of popular conversations. Location-aware technologies are new types of mobile technologies and to date very little has been written about how they influence our perception of public spaces and relationships to other people. Because they just recently became commercialized, most of the literature on locative media to date focuses on artistic and research projects. These projects, although relevant to understand the large-scale development of location-based applications today, are narrowly focused in the sense that they use specific applications and artworks as case studies (Admiraal, Akkerman, Huizenga, & Zeijts, 2009; Benford et al., 2004; Benford, Giannachi, Koleva, & Rodden, 2009; Crabtree et al., 2004; Flintham et al., 2003; Licoppe & Guillot, 2006; Licoppe & Inada, 2006). Now that location-aware mobile technologies have begun to be used by the general public, they have become an important piece in the construction of public spaces. Consequently, the issues we raise in this book are important for understanding and conceptualizing past, present, and future mobile technology use.

Filters and Control Mechanisms

The structure of our argument in this book is organized to show that location-aware technologies (and the desire to use them to interact with public spaces) did not emerge from nowhere. They belong to an ecology of other mobile technologies that also influenced how people interacted with and perceived public spaces. Many of the issues raised with location-aware technologies today have been experienced in similar forms in the past. Particularly, claims that mobile devices disconnect people from public spaces and cause them to not socially interact with people in their surroundings were already prominent since the development of earlier personal mobile technologies, such as the paperback book in the nineteenth century.

In Chapter 1, we argue that instead of disconnecting people from their surroundings, mobile technologies actually work as interfaces to public spaces, helping people control and manage their interactions with their surroundings. We show that the desire to selectively interact with public spaces comes from an inner desire to filter external stimuli in public spaces, named by Simmel (1950) the *blasé* attitude. This chapter discusses the growth of the modern city and how it affected people's interactions with public spaces, and the consequent ways people have longed to manage and control their experiences of urban spaces. This argument will contrast with the popular notion that mobile phones, the Walkman, and iPods "remove" individuals from the physical space they inhabit. For example, Michael Bull (2001) suggested that the Walkman had a major effect on people's perception of their surroundings by withdrawing them from their surrounding space. Bull argued that Walkman use allowed individuals to partially ignore the space they traveled through by imparting a personalized "soundscape" on the public space. Contrary to this view, we argue that individuals have always used mobile technologies as ways of framing their interactions with their surrounding space and paying selected attention to that space, but not necessarily through a withdrawal from it.

Using mobile technologies in public spaces has often been perceived as an invasion of the public by the private, and the divide between public and private spaces has always been problematic. Notably, it is also a divide that shifts as new technologies are introduced into the social fabric. In Chapter 2, we analyze how mobile technologies have always challenged the borders between what was public and what was private. We operate from a viewpoint that recognizes that neither public nor private are objective entities: They are socially negotiated and constantly shifting. We show how the ideas of public and private have assumed different meanings throughout history and employ a multifaceted approach to discuss how mobile technologies—namely, the portable camera, the book, the Walkman, and the iPod—complicate commonly understood divisions between private and public experience and make the socially negotiated nature of public and private visible.

The mobile phone has received more attention than any other mobile technology when it comes to discussing the borders between public and private spaces. Claims of disruption of public spaces by private conversations have been frequent in media discourses and in scholarly works (Katz & Aahkus, 2002; Ling, 2004, 2008; Plant, 2001). With mobile phones, private conversations were claimed to enter public spaces, disrupting the shared experience of those spaces. In a sense, mobile phones were claimed to privatize public spaces when people engaged in personal conversations with distant others. This disruption of the public by the private has led to accounts of the individualizing tendencies of mobile phones and has resulted in policies and social norms that dictate when individuals should engage in personal phone calls. In Chapter 3 we frame the existing literature on mobile phone use in public spaces within our earlier discussion of control in public spaces to show how these technologies also operate as sorts of selective interfaces to public spaces. People use mobile phones to actively choose when and where to engage with their physical surroundings; they do not necessarily use them to remove themselves from the physical setting. While mobile phones have problematized the divide between public and private, the inclusion of location-awareness in these mobile communication devices further complicates perceptions of place and location. After addressing how mobile phones were perceived as disconnecting users from place, we outline a brief history of location-aware technologies through locative media art and location-based services (LBS) to show how location-awareness brings location to the forefront of users' interactions with information and other users. Chapter 3 also explores how the emergence of location-aware mobile technologies not only redefines people's connections to locations, but also fundamentally redefines their meaning. We show that the ability to attach information to locations is a critical factor that differentiates location-aware technologies from other types of mobile technologies. While users might feel more empowered by their ability to "browse" public spaces, they are also exposed to privacy concerns, surveillance, and power differentials.

While the first section of this book develops a conceptual and historical framework for understanding how and why individuals use mobile technologies in public spaces, the second section explores the social and political implications of using location-aware technologies in public spaces. Specifically, we focus on three main issues that are weaved together by the idea of control: privacy, power, and the meaning of public spaces.

Chapter 4 deals with the privacy concerns raised by location-aware technologies, namely, the fear of losing control over one's location information—a concept which has been called *locational privacy* (Blumberg & Eckersley, 2009; Monmonier, 2002; Strandburg & Raicu, 2006). We highlight how locational privacy is different from privacy issues already addressed with the World Wide Web. When location-aware devices become the interface to access both the Internet and public spaces, existing issues of digital privacy acquire an additional element: location. These privacy concerns will also shape the development and implementation of LBS. We discuss two types of fears that have been articulated by users, designers, and the popular press: fear of top-down surveillance (mostly from governments and corporations), and fears of collateral surveillance (primarily from other people). Location information is becoming a key type of aggregated data collected by marketing companies in order to understand consumer behavior. Although this data is generally anonymous, users are often not aware of these practices. Having these concerns and practices in mind, we argue for a novel understanding of privacy in the context of location-aware technologies, an understanding which defines privacy in relation to users' control over their location information.

Control is not only related to privacy, but also to power. In Chapter 5, we explore the power relationships embedded in location-aware mobile technology use in public spaces. Location-aware technologies are elements of power networks that have the potential to discipline people's movements through space and the social relationships in it. In this chapter, we examine the power relationships mediated by the use of location-aware mobile interfaces. We address forms of power asymmetries in situations of collateral surveillance with location-aware technologies in different interpersonal contexts: parolees and parole officers, children and parents, and friends in location-based social networks (LBSNs). We then explore how location-aware technologies mediate the relationship between users and spaces. Important to our analysis is the awareness that location-based advertising (LBA) not only empowers individuals in public, but also corporations, who exert power over individuals who use location-based services (LBS). Finally, we explore how the use of location-aware interfaces, by allowing individuals to directly filter the information they access from locations, leads to different forms of social exclusion and fragmented perceptions of public spaces.

Lastly, Chapter 6 explores how location becomes an important piece of personal and spatial identity construction. We first explain how location is becoming another element of self-presentation in location-based social networks. Where one is becomes a fundamental part of who one is. By choosing to check-in to some places and not others, LBSN participants show their social network some aspects of their lives and not others. Those locations, then, become part of how others infer qualities about people. We then address the role of location-aware technologies in enabling users to read and write space by accessing location-based information left by others or by actively attaching information to locations (what we call *the presentation of location*). Because people can only access certain types of information when they are standing at the location where the information belongs, these technologies encourage specific spatial trajectories (Benford & Giannachi, 2011; Benford et al., 2009). These spatial trajectories are, however, different from early locative media art audio-walks and location-based art performances: They are created by common users, rather than by artists. As such, there is an increasing amount of location-based information that is constantly being created, erased, moved, and accessed that becomes an intrinsic part of a new dynamic urban landscape. Ultimately, this chapter is also about two forms of control: The ability to control the presentation of self through location, and the ability to control spaces through personalization and narrative construction.

The narrative in this book is weaved by another narrative: The story of Johanna, a woman in her mid-twenties who moves abroad to pursue a graduate degree. By following her first week in a foreign country, we are able to see how she users her personal mobile technologies to manage her own interactions with unfamiliar public spaces, and also how she perceives other people's uses of the same technologies. Johanna's experiences are not unique, and could be part of the life of anybody who owns and uses mobile technologies in public spaces. Some readers might identify with her experiences while others will not, but the situations she experiences are a common part of any urban landscape today. We hope that you can identify with her experiences while reading them, as we did while writing this book.

Notes

1   Turkle (1995) describes the culture of calculations as the computer culture in the 1960s and 1970s in which people (mostly computer scientists, hackers, and hobbyists) were required to know a programming language in order to interact with computers. By contrast, the culture of simulations grew out of the development of the Macintosh GUI in the 1980s, when non-expert "users" could intuitively manipulate icons that simulated a desktop on the computer screen.

2   However, Johnson (1997) warns us that "a printing press or a camera deals with representations as end-products or results; these machines are representational in that they print words on paper or record images on film, but the underlying processes are purely mechanical in nature. A computer, on the other hand, is a symbolic system from the ground up. Those pulses of electricity are symbols that stand in for zeros and ones,

which in turn represent simple mathematical instruction sets, which in turn represent words or images, spreadsheets or e-mail messages. The enormous power of the modern digital computer depends on this capacity for self-representation" (p. 15).

3   There are other important mobile technologies, including the transistor radio, and the BoomBox. However, we will not be addressing those technologies because they are not individual technologies. The transistor radio made music portable 30 years before the Walkman. Both the transistor radio, and the later BoomBox are mobile technologies, but they are not so much personal forms of media as they are collective forms. Whereas the iPod and the Walkman allow individuals to personalize their experience of place by imparting an individualized soundscape, the transistor radio and the BoomBox impart a soundscape that changes the auditory nature of the place for everyone. They colonize a place's soundscape for everyone in the vicinity. Our focus is instead on personal mobile media, so we have consequently limited our discussion to the book, the Walkman, the iPod, the mobile phone, and location-aware technologies.

4   *Foursquare* is a location-based social network application that includes gaming elements, such as badges and points. Every time a user "checks in" at a location, she accrues points, which eventually lead to her becoming the mayor of that location. Players compete with each other for mayorships and badges. Chapter 3 contains more information on *Foursquare*.

5   This definition of space as a place without meaning is clearly contested by many scholars, most notably by Henri Lefebvre's (1991) triadic logic of social space. Lefebvre coined the term "production of space" to highlight that space is composed of more than just its physical characteristics, namely a socially produced space. David Harvey (1993) suggested that "the strength of the Lefebvrian construction . . . is precisely that it refuses to see materiality, representation and imagination as separate worlds and that it denies the particular privileging of any one realm over the other, while simultaneously insisting that it is only in social practices of daily life that the ultimate significance of all forms of activity is registered. It permits, therefore, an examination of the processes of place construction in which . . . [it] owes as much to activities in the representational and symbolic realms as to material activities" (p. 23).

6   *The Gold Horns Thief* is a location-based game that includes challenges and missions, developed on the application SCVGR by two students at the IT University of Copenhagen, Louise McHenry and Marlene Ahrens in 2011. The students developed this game in association with the National Museum in Copenhagen with the intent to raise people's awareness about this important historical fact in the Danish history, and to motivate a younger generation to visit museum spaces.

7   Following Gilles Deleuze's (1994) ideas of actualization and virtualization.

References

Admiraal, W., Akkerman, S., Huizenga, J., & Zeijts, H. V. (2009). Location-based technology and game-based learning in secondary education: Learning about Medieval Amsterdam. In A. De Souza e Silva, & D. Sutko (Eds.), *Digital cityscapes: Merging digital and urban playspaces* (pp. 302–320). New York: Peter Lang.

Agnew, J. A. (1987). *Place and politics: The geographical mediation of state and society*. Boston: Allen & Unwin.

Agnew, J. A., & Duncan, J. S. (1989). *The power of place: Bringing together geographical and sociological imaginations*. Boston: Unwin Hyman.

Barkhuus, L., & Dey, A. (2003). Location-based services for mobile telephony: A study of users' privacy concerns. Paper presented at the Proceedings of the INTERACT 2003, 9TH IFIP TC13 International Conference on Human-Computer Interaction.

Benford, S., Flintham, M., Drozd, A., Anastasi, R., Dowland, D., Tandavanitj, N., et al. (2004). Uncle Roy all around you: Implicating the city in a location-based performance. Paper presented at the Proceedings of the Advanced Computer Entertainment Conference, Singapore.

Benford, S., & Giannachi, G. (2011). *Performing mixed reality*. Cambridge, MA: MIT Press.

Benford, S., Giannachi, G., Koleva, B., & Rodden, T. (2009). From interaction to trajectories: Designing coherent journeys through user experiences. Paper presented at the Proceedings of the 27th International Conference on Human Factors in Computing Systems.

Blumberg, A. J., & Eckersley, P. (2009). On locational privacy and how to avoid losing it forever. Retrieved from http://www.eff.org/wp/locational-privacy.

Boesen, J., Rode, J., & Mancini, C. (2010). The domestic panopticon: Location tracking in families. Paper presented at the UbiComp 2010.

Bull, M. (2000). *Sounding out the city: Personal stereos and the management of everyday life*. Oxford: Berg.

Bull, M. (2001). The world according to sound: Investigating the world of Walkman users. *New Media and Society, 3*, 179–197.

Bull, M. (2007). *Sound moves: iPod culture and urban experience*. New York: Routledge.

Casey, E. (2009). *Getting back into place: Toward a renewed understanding of the place-world* (2nd ed.). Bloomington, IN: Indiana University Press.

Castells, M. (2000). *The rise of the network society*. Oxford: Blackwell.

Consolvo, S., Jung, J., Greenstein, B., Powledge, P., Maganis, G., & Avrahami, D. (2010). The wi-fi privacy ticker: Improving awareness & control of personal information exposure on wi-fi. Paper presented at the UbiComp '10. Retrieved from http://seattle.intel-research.net/people/daniel/pubs/Consolvo_UbiComp_10.pdf.

Consolvo, S., Smith, I. E., Matthews, T., LaMarca, A., Tabert, J., & Powledge, P. (2005). Location disclosure to social relations: Why, when, & what people want to share. Paper presented at the Proceedings of the SIGCHI Conference on Human Factors in Computing Systems.

Crabtree, A., Benford, S., Rodden, T., Greenhalgh, C., Flintham, M., Anastasi, R., et al. (2004). Orchestrating a mixed reality game "on the ground." Paper presented at the Proceedings of the SIGCHI Conference on Human Factors in Computing Systems.

Cresswell, T. (2004). *Place: A short introduction*. Malden, MA: Blackwell.

De Certeau, M. (1988). *The practice of everyday life*. Minnesota: University of Minnesota Press.

de Souza e Silva, A. (2006). From cyber to hybrid: Mobile technologies as interfaces of hybrid spaces. *Space and Culture, 3*, 261–278.

de Souza e Silva, A. (2009). Hybrid reality and location-based gaming: Redefining mobility and game spaces in urban environments. *Simulation & Gaming, 40*(3), 404–424.

de Souza e Silva, A., & Frith, J. (2010). Locational privacy in public spaces: Media discourses on location-aware mobile technologies. *Communication, Culture & Critique, 3*(4), 503–525.

de Souza e Silva, A., & Sutko, D. M. (2009). *Digital cityscapes: Merging digital and urban playspaces*. New York: Peter Lang.

de Souza e Silva, A., & Sutko, D. M. (2011). Theorizing locative technologies through philosophies of the virtual. *Communication Theory, 21*(1), 23–42.

Deleuze, G. (1994). *Difference and repetition*. New York: Columbia University Press.

Devitt, K., & Roker, D. (2009). The role of mobile phones in family communication. *Children & Society, 23*(3), 189–202.

du Gay, P., Hall, S., Janes, L., Mackay, H., & Negus, K. (1997). *Doing cultural studies: The story of the Sony Walkman*. London: Sage.

Flintham, M., Benford, S., Anastasi, R., Hemmings, T., Crabtree, A., Greenhalgh, C., et al. (2003). Where on-line meets on the streets: Experiences with mobile mixed reality games. Paper presented at the Proceedings of the SIGCHI Conference on Human Factors in Computing Systems.

Gane, N., & Beer, D. (2008). *New media: The key concepts*. New York: Berg.

Gergen, K. (2002). The challenge of absent presence. In J. Katz, & M. Aakhus (Eds.), *Perpetual contact: Mobile communication, private talk, public performance* (pp. 227–241). New York: Cambridge University Press.

Geser, H. (2004). Towards a sociological theory of the mobile phone. *Swiss Online Publications*. Retrieved from http://socio.ch/mobile/t_geser1.htm.

Gordon, E. (2008). Towards a theory of network locality. *First Monday*, *13*(10).

Hansen, J., Alapetite, A., Andersen, H., Malmborg, L., & Thommesen, J. (2009). Location-based services and privacy in airports. In T. Gross, J. Gulliksen, P. Kotzé, L. Oestreicher, P. Palanque, R. Prates, & M. Winckler (Eds.), *Human-computer interaction – INTERACT 2009* (vol. 5726, pp. 168–181). Berlin, Heidelberg: Springer.

Harrison, S., & Dourish, P. (1996). Re-place-ing space: The roles of place and space in collaborative systems. Paper presented at the Proceedings of the 1996 ACM Conference on Computer Supported Cooperative Work.

Harvey, D. (1993). From space to place and back again. In J. Bird, B. Curtis, T. Putnam, & G. Robertson (Eds.), *Mapping the futures: Local cultures, global change* (pp. 3–29). London: Routledge.

Harvey, D. (1996). *Justice, nature and the geography of difference*. Cambridge, MA: Blackwell Publishers.

Hong, J. I., Borriello, G., Landay, J. A., McDonald, D. W., Schilit, B. N., & Tygar, J. D. (2003). Privacy and security in the location-enhanced world wide web. Paper presented at the Proceedings of the Workshop on Privacy at Ubicomp.

Hosokawa, S. (1984). The Walkman Effect. *Popular Music 4*, 165–180.

Humphreys, L. (2007). Mobile social networks and social practice: A case study of Dodgeball. *Journal of Computer-Mediated Communication*, *13*(1), article 17.

Ito, M., Okabe, D., & Matsuda, M. (Eds.). (2005). *Personal, portable, pedestrian: Mobile phones in Japanese life*. Cambridge, MA: The MIT Press.

Johnson, S. (1997). *Interface culture: How new technology transforms the way we create and communicate*. San Francisco, CA: Harper Edge.

Katz, J., & Aahkus, M. (2002). *Perpetual contact: Mobile communication, private talk, public performance*. Cambridge: Cambridge University Press.

Lefebvre, H. (1991). *The Production of Space*. Malden, MA: Blackwell Publishers.

Lévy, P. (2001). *Cyberculture*. Minneapolis: University of Minnesota Press.

Lévy, P. (2004). *As Tecnologias da Inteligência: O Futuro do Pensamento na Era da Informática*. São Paulo: Editora 34.

Licoppe, C., & Guillot, R. (2006). ICTs and the engineering of encounters. A case study of the development of a mobile game based on the geolocation of terminals. In J. Urry, & M. Sheller (Eds.), *Mobile technologies of the city* (pp. 152–163). New York: Routledge.

Licoppe, C., & Inada, Y. (2006). Emergent uses of a multiplayer location-aware mobile game: The interactional consequences of mediated encounters. *Mobilities*, *1*(1), 39–61.

Ling, R. (2004). *The mobile connection: The cell phone's impact on society*. San Francisco: Morgan Kaufman.

Ling, R. (2008). *New tech new ties*. Boston: MIT Press.

Ling, R., & Yttri, B. (2002). Hyper-coordination via mobile phones in Norway. In J. Katz, & M. Aakhus (Eds.), *Perpetual contact: Mobile communication, private talk, public performance* (pp. 139–169). New York: Cambridge University Press.

Manovich, L. (2002). *The language of new media*. Cambridge, MA: MIT Press.

Massey, D. (1993). Power-geometry and a progressive sense of place. In J. Bird, B. Curtis, T. Putnam, G. Robertson, & L. Tickner (Eds.), *Mapping the futures: Local cultures, global change* (pp. 56–69). London, New York: Routledge.

Massey, D. (1995). The conceptualization of place. In D. Massey, & P. Jess (Eds.), *A place in the world? Places, cultures and globalization* (pp. 45–77). Oxford: Oxford University Press.

Monmonier, M. S. (2002). *Spying with maps: Surveillance technologies and the future of privacy*. Chicago: University of Chicago Press.

Moores, S. (2004). The doubling of place: Electronic media, time-space arrangements and social relationships. In B. Couldry, & A. McCarthy (Eds.), *Media/Space: Place, scale and culture in a media age* (p. 21). London: Routledge Comedia Series.

Peters, J. D. (1999). *Speaking into the air: A history of the idea of communication*. Chicago: University of Chicago Press.

Plant, S. (2001). *On the mobile. The effects of mobile telephones on social and individual life*. London: Motorola Inc.

Puro, J. P. (2002). Finland, a mobile culture. In J. Katz, & M. Aakhus (Eds.), *Perpetual contact: Mobile communication, private talk, public performance* (pp. 19–29). Cambridge: Cambridge University Press.

Rheingold, H. (1993). *The virtual community: Homestead on the Electronic Frontier*. Cambridge, MA: The MIT Press.

Sack, R. D. (1992). *Place, modernity, and the consumer's world: A relational framework for geographical analysis*. Baltimore, MD: Johns Hopkins University Press.

Schivelbusch, W. (1986). *The railway journey: The industrialization of time and space in the 19th century*. Berkeley, CA: University of California Press.

Shklovski, I., Vertesi, J., Troshynski, E., & Dourish, P. (2009). The commodification of location: Dynamics of power in location-based systems. Paper presented at the Proceedings of the 11th International Conference on Ubiquitous Computing.

Simmel, G. (1950). *The sociology of Georg Simmel* (K. Wolff, Trans.). New York: Free Press.

Simpson, J. (1999). *Oxford English dictionary*. Oxford: Oxford University Press.

Smith, D. M. (2000). *Moral geographies: Ethics in a world of difference*. Edinburgh: Edinburgh University Press.

Strandburg, K. J., & Raicu, D. S. (2006). *Privacy and technologies of identity: A cross-disciplinary conversation*. New York: Springer Science+Business Media.

Tuan, Y.-f. (1977). *Space and place: The perspective of experience*. Minneapolis: University of Minnesota Press.

Turkle, S. (1995). *Life on the screen: Identity in the age of the Internet*. New York: Simon and Schuster.

Turkle, S. (2011). *Alone together: Why we expect more from technology and less from each other*. New York: Basic Books.

Verstraete, G., & Cresswell, T. (2002). *Mobilizing place, placing mobility: The politics of representation in a globalized world*. Amsterdam: Rodopi.

SECTION I

Mobile Interfaces in Public Spaces

1

INTERFACES TO PUBLIC SPACES

Johanna is waiting for her flight, sitting at her gate among a crowd of people. She never liked airports. Airports are weird places: They contain mostly people who are strangers to each other, and who are constantly on the move, coming and going. Johanna feels uncomfortable and intimidated by the strangers around her. The couple sitting behind her are arguing, not especially loudly, but loudly enough that she can hear every word they say. Johanna has been trying to read, but she decides to give up on her book because she cannot concentrate with everything going on around her. She puts her book down, reaches into her backpack and takes out her iPod. She then puts her headphones on and turns her music up so that she can stop paying attention to the couple's angry conversation.

The first song that comes through her headphones is a soft, acoustic song from one of her favorite songwriters. She loves that song, but the busy airport space doesn't quite match the song's slow-paced rhythm, so she changes to a faster song. Replacing the sounds of the airport with a familiar melody helps her feel more comfortable in that space. She spends the next ten minutes listening to music and watching strangers as they make their way past their gate. Soon she notes on the monitors above her head that it is time for her to board. She turns off the iPod, and walks to her gate.

When we experience a place, we do so through our body, which acts as a layer between a place and our perception of it. We also develop techniques to filter the information around us, further interfacing our experiences. Nowhere can this be seen more clearly than in public spaces and urban areas. The crowded streets of the city contain so much action, so much information, that we constantly enact ways to mentally filter those places, choosing one thing to focus on over another. In other words, we find new ways to manage attention (Benjamin, 1980 [1929]; Goldhaber, 2006; Lanham, 2006). Managing attention has become increasingly important not only with the growth of urban spaces, but also of digital spaces. Richard Lanham (2006) has claimed that attention is a scarce commodity in today's digital society. With the growth of the Internet and other communication channels, we are being overloaded with information. He writes that, "In an information society, the scarce commodity is not information—we are choking on that—but the human attention required to make sense of it" (Lanham, 1997, p. 164). Lanham (2006) later called this the new economics of attention, arguing that because there is so much stimulation in the information society, keeping viewers' attention takes precedence over everything else. Similarly, Goldhaber (2006) suggests that "in an attention economy, one is never not on, at least when one is awake, since one is nearly always paying, getting or seeking attention, in ways and modes that are increasingly organized and tend to involve ever-large and more dispersed audiences" (n.p.). In the digital age, people are accessing increasing amounts of information. They are stimulated in new and diverse ways, and therefore feel the need to develop new mechanisms to manage their attention and decide what to absorb from the world around them.

We are certainly not living through the first era to face what is perceived as information overload. Discussing the development of the printing press, Michael Hobart and Zachary Schiffman (1998) write that "printing gave individuals access to a previously unimaginable number of books, overloading them with diverse and contradictory information" (p. 89). People living in the sixteenth century responded by developing new ways to organize information and attention, most notably in the forms of lists such as bibliographies and indexes. The authors point out that lists are a definite sign of information overload, and the sixteenth century stands "as one of the great ages of list making" (Hobart & Schiffman, 1998, p. 104).

However, as Jonathan Crary (2001) points out, the idea of attention itself is a nineteenth-century construction. The process of modernization, and along with it the growth of the modern city, brought the problem of attention to the foreground of modern urban life. It was in the late nineteenth and early twentieth centuries that "inattention . . . began to be treated as a danger and a serious problem, even though it was often the very modernized arrangements of labor that produced inattention" (Crary, 2001, p. 13). As Crary observes, the problem of attention emerged from a social and urban field increasingly saturated with sensory input. Urbanites moved through city streets that had become increasingly crowded, increasingly filled with sensory stimulation. Writing about these early nineteenth- and twentieth-century urbanites, Georg Simmel (1950) argued that the metropolitan man [sic] had to develop new ways to deal with the sensory stimulation of the city. He claimed individuals did this by reconstituting their psychic state, reformatting their attention by, in a sense, fragmenting their space; in other words, they developed new ways to filter their experience of the city. Simmel called this development the blasé attitude, arguing that urban individuals developed "an organ" that helped them deal with the stimulation of the city. The organ represented a new, intellectually refined approach to urban space. As Simmel (1950) says, "He [the metropolitan man] reacts with his head instead of his heart" (p. 410). The metaphorical organ Simmel describes is basically a filter between the sensory stimulation of the city and the mind of the metropolitan individual. In other words, it was a medium—or, as we suggest, a type of mental interface—that helped individuals manage their interactions with the urban environment.

While Simmel discussed the mental attitude of the metropolitan individual as a way to filter urban spaces, people have often used mobile technologies to accomplish a similar same goal. The blasé attitude as a technique and mobile media as technologies both are used by urban individuals to interface with the urban environment and to manage their experiences. For this reason, mobile technologies work as interfaces to public spaces. Going as far back as the development of the habit of reading in public spaces (Manguel, 1997; Schivelbusch, 1986), people have used mobile interfaces—such as books and newspapers—to alter the way they perceive public spaces and how they manage their interactions with other people and things in those spaces. By reading a book on public transit, individuals can divert their attention away from their surroundings, focusing on the narrative of a novel rather than the stimulation of the place. With the development of auditory personal mobile technologies, especially the Walkman and the iPod, it became even easier for people to enact a technologically enabled filter that helped them choose how to engage with public space. These technologies have often been accused of diverting people's attention and causing them to ignore the physical space and the people around them. However, undivided attention (to people, to spaces) is an unachievable, idealized goal. In reality, our attention spans cycle through different things and environments— and mobile technologies help people manage these cycles.

It has been common for scholars to argue that mobile technology use "withdraws" people from public space (Gergen, 2002; Goldberg, 2007; Puro, 2002), an argument we address throughout this book.[1] In this chapter, we address this concern by showing that people do not just use these technologies to "withdraw" from space. Rather, people often use mobile technologies to accomplish a similar goal as the blasé attitude: interface their relationships with other people and the space around them. We use the examples of three types of personal mobile technologies—the book, the Walkman, and the iPod—to show

how they can be analyzed as interfaces that help people manage and control their interaction with the public spaces around them. In these first two chapters, we focus on these three monologic mobile technologies. From Chapter 3 on we focus on mobile phones and location-aware technologies as dialogic forms of mobile media, discussing how newer mobile technologies complicate these arguments. For now, we look at the book, the Walkman, and the iPod as a way to develop our argument about mobile technologies as interfaces to urban spaces. Whereas with Simmel's (1950) *blasé* attitude the metropolitan man "reacts with his head instead of his heart" (p. 410), when people move through space with a mobile technology such as the Walkman, they "react with their head(phones) instead of their heart."

This chapter examines urban sociability as it relates to the use of these mobile technologies and develops a conceptual framework for understanding how and why individuals use mobile technologies in public spaces. To do so, we first examine issues of urban sociability by discussing two interrelated areas: (1) the growth of the modern city and how it has affected urban sociability; and (2) the ways people have longed to control their experience of urban spaces. Rather than arguing that mobile technologies themselves create a need to exert control over heterogeneity in the city, we instead believe that people have always longed to control both the built environment and their interactions with other people in public. We then move on to analyze mobile technology use and argue that through books and auditory media, individuals are able to enact a more controlled form of engagement with the stimulation of urban spaces. Just like with Simmel's psychological filter, individuals turn toward mobile interfaces not as a way to completely distance themselves from the experience of the city, but rather to choreograph an economy of attention that simultaneously distances and re-approximates them from urban spaces.

The Growth of the Modern Metropolis and the Decline of Public Life

In 1938, Louis Wirth wrote that "the growth of cities and the urbanization of the world is one of the most impressive facts of modern times" (p. 2). Wirth's words still ring true more than 70 years later, and the process of urbanization has continued more or less unchecked through the first decade of the twenty-first century. The move from rural to urban changed the ways in which people organize their daily lives, and more recently, the move toward megalopolises has reorganized urban life even more noticeably. In Mexico City, for example, Néstor Garcia Canclini (2001) has shown that the enormous growth of cities has made it difficult for its citizens to even identify themselves as residents of the city; instead, they often identify with the section of the city in which they live because the city as a whole is simply too large to have much meaning as a place of residence. Similar situations occur in other megacities, such as São Paulo, Tokyo, and New

York. The growth and sprawl of the built environment has also changed the way in which people negotiate the urban. Canclini (2001) discusses how the fractured nature of megacities has led to a confusion of identity and a shift toward more a private experience because public spaces have become less and less effective as pleasant sites of social interaction. Faced with the growth of global metropolises, it becomes important to return to Wirth's proclamation and examine how the built environment of cities over the last three centuries has changed the way in which people engage with urban areas.

Cities underwent a massive transformation around the time of the Industrial Revolution. From 1595 to the middle of the eighteenth century, the population of London grew from 315,000 to 750,000. In the nineteenth century alone, following the growth of industrialization, the population of London grew to over five million people (Sennett, 1977, p. 50). Other major cities, such as Paris and New York followed a similar trend, and much of the migration to urban areas came from rural towns. The new generations of city dwellers were confronted with an unfamiliar situation: No longer did they know the people they passed by on a daily basis, and the shift from rural to urban demanded new forms of sociability. Erving Goffman (1963) describes those demands well in his formulation of the "nod line":

> In Anglo-American society there exists a kind of "nod line" that can be drawn at a particular point through a rank order of communities according to size. Any community below the line, and hence below a certain size, will subject its adults, whether acquainted or not, to mutual greetings; any community above the line will free all pairs of unacquainted persons from this obligation.
>
> (pp. 132–133)

With the rapid population growth of urban areas, the streets of the metropolis fell far above Goffman's nod line. In contrast, in rural communities and smaller cities, most people still know each other or at least know of each other. As Goffman describes, they acknowledge each other's presence because it would be rude not to do so and because it is likely they will run into each other again. However, with the increasing migration from rural to urban areas, for the first time most people lived in a place where they did not know the vast majority of people they saw daily. The public spaces of the city demanded that urban individuals develop new social techniques to manage the stimulation of city streets and to interact with strangers.

Ultimately, much more than in rural communities, the city is where we come into contact with strangers, and through that contact we ideally learn from others who do not share our backgrounds and experiences, or as Sennett (1992) puts it: "A city ought to be a school for learning how to lead a centered life. Through exposure to others, we might learn how to weigh what is important and what is not" (p. xiii). Before the growth of vast urban areas, the benefits offered by the private spaces of the home, tight-knit community in rural communities and smaller cities were more clearly demarcated from the benefits offered by the public spaces of the city, and each served its own purpose (Sennett, 1977). Private communities were places for strong emotional attachments; public places were where individuals interacted with others on a less personal level, but it was this less personal level that allowed for people to learn from others and expand past their tight-knit, typically homogenous groups.

Sennett's major argument in *The Fall of Public Man* (1977) is that the balance between private and public has suffered in the last century and a half of urban life, with the private coming to dominate the public. Sennett attributes this change to the growth of personality in the late nineteenth and twentieth centuries. Individuals turned inward because of a growing narcissism that valued personally meaningful interactions as the only interactions worth having. However, as Claude Fischer (2005) has argued, this turn away from the public and toward tight-knit, more private groups is not necessarily a growth in individualism; instead, it is a turn more toward social privatism. Fischer writes that what we observe is a desire to engage with "a more private world of family, work, and friends—a story of greater, but still social, privatism" (p. 6). It is not that people become hermits or turn away from society. Instead, they eschew the more uncontrolled interactions with strangers in the city for the more controlled interactions of family and close friends.

Whether we label it social privatism or individualism, the argument remains much the same. Over the last 200 years, the private began to encroach on the public, with individuals turning away from public spaces because the interactions with strangers could not fulfill their desires in the same way that private inter-actions could. The growth of the city was a partial cause of this change, as the more controlled public life of eighteenth-century cafes turned into the crowded city streets outlined most famously by Baudelaire's (1964) concept of the *flâneur*. Individuals came to value the emotional attachments of community and the home at the expense of the impersonal, ephemeral attachments with other individuals in public places. In a way, this shift in urban life can be conceptualized as a restructuring of the choreography of attention: To manage the stimulation of crowded metropolises, individuals turned inward and avoided connecting with the people they passed by every day.

Besides Sennett and Simmel, other scholars also focused on the city as a space of heterogeneity (Canclini, 2001; De Certeau, 1988; Jacobs, 1961). Jane Jacobs (1961), for example, writes that "Great cities are not like towns, only larger. They are not like suburbs, only denser. They differ from towns and suburbs in basic ways, and one of these is that cities are, by definition, full of strangers. To any one person, strangers are far more common in big cities than acquaintances" (p. 30). In addition, Néstor Garcia Canclini (2001) notes that the modern city became a space of increasing intercultural segmentation and hybridization, with the co-existence of different types of people and architecture that converge in public spaces. Heterogeneity was also one of the defining characteristics of the city according to the Chicago School of Sociology, most notably in the work of Louis Wirth (1938). For all of these thinkers, city life is a life of co-existence with other people one does not know and who one has little in common with. It is a life of heterogeneity that stands in contrast to the homogeneity (though of course not complete) of rural life, in which people know one another and often have a great deal in common.

Heterogeneity may be a defining element of urban life, but it is not always pleasant. People enjoy suburbs for a reason; they offer a more stable, sometimes safer experience of homogeneity. In contrast, heterogeneity can be exciting and stimulating, but it can also be overwhelming and frightening. As a result of this desire to control heterogeneity, modernist city planners sought to control the heterogeneity and randomness of urban spaces through zoning laws and architectural practices that created isolated urban enclaves seemingly separate from the city as a whole. For example, many city neighborhoods are built and zoned following a single-use model in which housing is not mixed with shops and businesses. This reduces the diversity of people one can find in a space and has led to a decline in what was formerly perceived as vibrant and social public spaces because public spaces were often transformed into business districts with reduced pedestrian traffic. As Jane Jacobs (1961) argued, rationalist city planning valued order and efficiency over randomness and vibrancy.

The rationalist urban planning model, exemplified most notably by the city plans of Robert Moses (Caro, 1975), were supposedly a response to the growing stimulation of urban spaces. These plans, which included large-scale highway projects such as the Brooklyn–Queens expressway, attempted to control the built environment by imposing a new sort of functionality on the city. This new functionality, however, worked to the detriment of livable, vibrant public spaces. It was this turn away from usable, accessible public spaces that William Whyte (1980) and his followers began to critique in the 1970s. Whyte examined why so many public spaces in the modern city failed, outlining methods for building public spaces in which people actually wanted to go to. Ultimately, Whyte's argument was that most urban spaces were not conducive to public life; most were not spaces that people went to interact with strangers. People instead valued mobility over sociability, and homogeneity over heterogeneity, moving through public spaces mostly as a way to make it back to the more private, controlled spaces of the home or close friends.

The construction of high rises, skywalks, and the infrastructure of automobility has led to "dead public spaces," where the built environment allows individuals to move freely without having to spend much time traversing spaces crowded with strangers (Graham & Marvin, 2001; Sennett, 1977, 1992). Of all these urban infrastructures, the one that has likely done the most damage to public space is the infrastructure of automobility (Packer, 2008; Urry, 2000, 2007). In an interview with Jelle Bouwhuis, Norman Klein notes how the construction of highways across the United States in the 1950s and 1960s led to the creation of what he calls the "city of circulation" (Bouwhuis, 2007), which ultimately contributed to the erosion of urban centers and the justification for their renewal. In the city of circulation, physical space is mere transit space. Transit space has the potential to destroy the crowded vibrancy of good public spaces where people stop to share the experience of the city with others. Automobility also enabled the growth of the suburbs and with that growth, the flows of urban life have often moved from the center of the city to the periphery of the suburbs (Scott & Soja, 1998; Soja, 1996).

Although many urbanists have tried vigorously to reproduce what Whyte (1980) described as "good public space" by designing pedestrian intersections and street furniture to produce public spaces where strangers would congregate (Schectman, 2009), it remains true that many contemporary cities are not built to facilitate interactions between strangers. Sennett (1992) traces this trend to what he calls, playing off Max Weber (2003), the "protestant ethic of space":

> This compulsive neutralizing of the environment is rooted in part in an old unhappiness, the fear of pleasure, which led people to treat their surroundings as neutrally as possible. The modern urbanist is in the grip of a Protestant ethic of space.
>
> (p. 42)

The neutralized environment is reflected in the strip malls that dominate the periphery of most major urban areas as well as downtown areas that have been transformed from places of public life into financial districts that are practically empty outside of work hours. These built environments value the functionalism of mobility over the values of random sociability. Through an emphasis on automobiles over pedestrians, people are able to drive through areas without having to socialize with other people. We can see this in any urban or suburban area where high-speed roads replace sidewalks. Architectural movements such as New Urbanism have arisen to attempt to re-imagine how we should build urban and suburban spaces to provide more pedestrian friendly built environments (Schectman, 2009). New Urbanism is an architectural design movement that arose in the mid-1980s and promotes development of walkable neighborhoods and mixed-use development.[2] New Urbanists, such as Michael E. Arth and Leon Krier, often focus on walkability and design areas in ways that were popular before the rise of the automobile, emphasizing sidewalks and city centers. That is markedly different from most contemporary urban and suburban spaces that emphasize automobiles over pedestrians. If (auto)mobility is valued over any other kind of connection to urban spaces, then individuals are able to effectively avoid interacting with the spaces they move through or the people who occupy those spaces.

The built environment undoubtedly affects the way in which people socialize in urban areas. It is too simple, however, to fall back on spatial determinism. Population growth and infrastructures of mobility do not fully determine public life in the city; instead, the built environment exists in a reflexive relationship to other societal factors that influence public life. While the decrease in random sociability is a consequence of the growth of rationally planned urban centers, people have rarely interacted with strangers in cities. Heterogeneity has been considered a positive value in planning and academic circles, but people still tend to seek out homogeneity. The built environment does partially determine how people live, but the way people live also shapes the built environment.

Stimulation and an Individualized Experience of Space

One of Simmel's (1950) main arguments in "The metropolis and mental life" is that individuals must struggle to maintain their individuality in the face of the growing material culture of the city. Because individuals were made to feel negligible when faced with busy city streets and the growing money economy, they longed for a way to individualize their experience of the city. As Eric Gordon and Adriana de Souza e Silva (2011) show, drawing from Simmel (1950), "The city was incomprehensible in its unfiltered form, so having mental reserve was required to parse out the various social situations from the sites and sounds of the urban street" (p. 87). People did so, according to Simmel, by developing a more intellectual, refined approach to urban spaces. Rather than interacting with space as a totality, metropolitan individuals compartmentalized their perception of urban space. The development of this intellectual approach to the metropolis worked as a type of mental interface, a way to filter the information present in a place. Describing this phenomenon, Simmel (1950) writes,

> The metropolitan type of man—which of course, exists in a thousand individual variants—develops an *organ* protecting him against the threatening currents and discrepancies of his external environment which would uproot him.
>
> (p. 410, emphasis added)

Simmel describes this metaphorical organ as the development of the *blasé* attitude, which "results first from the rapidly changing and closely compressed contrasting stimulations of the nerves" (p. 414). Each person, each sound, each building, could not be given full attention as an individual entity; to do so would make passage on city streets impossible. Instead, perception of urban space became blunted, and no "one object deserves preference over the other" (p. 415). The development of the *blasé* attitude allowed people to filter the way they perceived the city by developing an intellectual approach that denies the individuality of objects and people. If they did not develop this attitude, and instead interacted with people and objects as they did in rural communities, where there were less people to interact with and most people knew each other, urban individuals would be completely overwhelmed. So, reframing Simmel's perspective, we suggest that the *blasé* attitude is an early type of mental interface individuals developed to manage and filter their interactions with the growing urban spaces. As we will see later, this psychological interface is often complemented by other types of technological interfaces, such as books and newspapers.

Paradoxically, the blunting of discrimination that led the individual to homogenize the objects of the city also contributed to a new form of subjectivity, what Simmel calls the "individualization of mental and psychic traits" (p. 420). The picture Simmel paints of the early twentieth-century city is of a small individual standing in a street crowded with strangers, overwhelmed by material culture, and made insignificant by the flows of the money economy. It is a picture to which many people can certainly relate, and as discussed in the previous section, a picture that could still be painted over 100 years later. Interestingly, when metropolitan individuals were made to feel small and negligible, they responded by carving out their own individualized experience. As Simmel describes it, "the individualization of mental and psychic traits" resulted in urbanites seeking to differentiate themselves from the masses by not participating fully in the crowd or by doing things, such as wearing flamboyant clothing, so that they would stand out from the crowd. Fighting against the nearly overwhelming growth of the city and the money economy, urban individuals turned inward as a way to differentiate themselves from the material world around them, "summoning the utmost in uniqueness and particularization, in order to preserve his most personal core" (p. 422).

We can see in Simmel's description of the metropolitan man's [sic] urge for an individualized experience of the city an initial gesture toward the desire to control the spaces around them, a desire which has been increasingly fulfilled through the use of mobile technologies in public spaces. Michael Bull (2000), for example, writes that Walkman users impart a personalized, unique soundscape on their experience of movement and public place. The desire for this personalized experience comes from what Simmel identifies as the need to summon the "utmost in uniqueness and particularization" in the face of increasingly stimulating urban spaces.

As discussed above, the ever-increasing need to reaffirm a personal core is what Sennett (1977) saw as the contributing factor in the decline of public life. As we have seen, Sennett contrasts public life—"the place where strangers meet"—with the private life of families and insular communities. He argues that individuals have turned inward, embracing a "cult of personality" that has severely damaged public life. Unlike Simmel, Sennett argues that the turn inward, the need for a unique experience of the city, is caused by a turn away from the rational, calculating mind and toward an affective approach to social interactions. Whereas Simmel saw the objective spirit of the city—that is, the built environment, the flows of the money economy, and the laws and rules of the city—overcoming the subjective spirit, Sennett sees private life encroaching on public life. Other scholars, notably Robert Putnam (2000), have made similar claims about a turn inward away from public life. The overvaluation of private life results in urban individuals bringing the same expectations they have in their private lives out into the public streets. Because urban dwellers cannot hope to connect with strangers at the same intimate level as they do with their strong ties, they are more likely to long for an individualized, controlled experience of the city and withdraw from the random sociability with strangers that long defined public life (Fischer, 2005; Putnam, 2000).[3]

While Sennett and Simmel take a more sweeping approach to analyzing how individuals control their experiences of urban life, Goffman's (1963) close analysis of micro-level interactions can also provide fruitful grounds for understanding why and how people take steps to exert control over sociability. Goffman discusses the risks associated with welcoming face-to-face interactions and associating with strangers. There are undoubtedly benefits to welcoming contact with others, including the expansion of horizons to which Sennett points, but there are also serious downsides. By opening oneself to others, one "opens himself up to pleadings, commands, threats, insult and false information" (p. 105). These are all practical reasons why urban individuals strive for an individualized experience of the city in Simmel's terms, or seek only close relationships that can be better controlled. Closing off contact with strangers reduces the danger of urban life. There are many ways one can close off contact with strangers, ranging from micro-level behaviors such as body positions or facial expressions to the use of mobile technologies such as the book or the iPod, and mobile phones as we will see in the next section and following chapters, respectively.

While the risk associated with opening oneself up to strangers certainly leads to a more withdrawn experience of public space, the major reason people do not initiate conversation with strangers is out of a different type of fear. As Goffman (1963) puts it,

> The force that keeps people in their communication place in our middle class society seems to be the fear of being thought forward and pushy, or odd, the fear of forcing a relationship where none is desired—the fear in the last analysis, of being rather patently rejected and even cut.
>
> (p. 140)

As we isolate ourselves, that isolation becomes a self-perpetuating force affecting sociability in public places. The valuation of emotional connection in the city makes it even more difficult to then initiate communication with strangers; the fear of being rejected is too strong. The impersonal contact of the street has decreased, making it more difficult for public life to be re-established because of the great likelihood of rejection. As we outlined earlier, this tendency is reflected in the built environment of many urban areas, a built environment that lacks the necessary characteristics for facilitating random sociability.

The growth of individuality has affected urban spaces by de-emphasizing communication among strangers and forcing individuals to create mechanisms to deal with the uncomfortable situation of being among strangers. We can control our socialization with strangers in many ways, whether through a mental reserve, the building of privatized infrastructure that takes people off the street, or the growth of insular communities. But the most visible, and possibly the most overlooked, way we interface our interactions in urban spaces (and the strangers in them) is through the use of mobile technologies. While today dialogic mobile phones and location-aware devices have been claimed to individualize space and lead to social atomization (Hampton, Livio, & Sessions, 2010; Humphreys, 2007; Wellman, 2002), we can see in examples of other monologic mobile technologies how people tactically use them to interface their experience of space. Before mobile phones, the book, the Walkman (and now the iPod) already worked as techno-logical interfaces that enable people to control their experience of urban spaces.

Mobile Technologies as Interfaces to Urban Spaces

Mobile technologies, including the Walkman and the mobile phone, have often been identified as ways in which individuals can "withdraw" from public spaces and lose connection with their surroundings (Bull, 2000; Gergen, 2002; Hosokawa, 1984; Plant, 2001). This argument makes some sense. People listening to headphones on a bus do, in a way, withdraw from the shared space of the bus and inhabit an imaginary world of elsewhere. They find shelter from the shared space by imparting their own soundtrack on their experience of the bus. But this situation can also be viewed through a different lens. Instead of analyzing mobile technology use in public spaces through the ideas of withdrawal and disconnection, we can look at these technologies as interfaces to public spaces. This approach enables us to more accurately understand why and how individuals use mobile technologies in public spaces.

The book was one of the first types of mobile technologies, but books were not always mobile. Early Mesopotamian books (around the third millennium B.C.) consisted of a set of clay tablets that could be held in hand but not easily transported. The same applies to the papyrus scrolls that replaced clay tablets in Ancient Egypt in the second millennium B.C. Even after codex books began to replace classical scrolls in the first century, their portability was limited. During the Middle Ages, most manuscripts were designed to be read at libraries, not to be transported. The invention of reading machines, such as Agostino Ramelli's "rotary reading desk" in 1588 (Figure 1.1) demonstrates this lack of mobility.

Books did not decrease in size until roughly 100 years after the invention of Gutenberg's printing press. In the late sixteenth century, students began using hornbooks, which were small enough to fit in students' hands; however, books

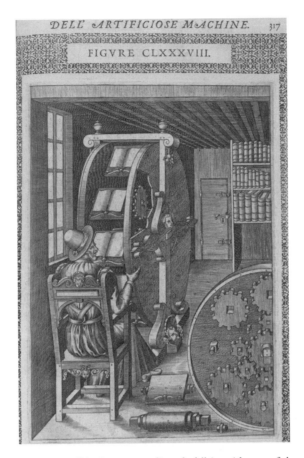

DELL' ARTIFICIOSE MACHINE.   317

FIGVRE CLXXXVIII.

Figure 1.1 Agostino Ramelli's "rotary reading desk" is evidence of the non-portability of books until at least the sixteenth century. Illustration from *c.*1588. Copyright: Elizabethan Club of Yale University.

did not take their current form until the Industrial Age. The nineteenth century saw the first printing of the pocket paperback, which established the book as the dominant form of mobile media. The contributing factor in the move to the pocket paperback was not technical improvements to printing technologies; rather, social forces played the major role in shaping the new mobility of the book. Alberto Manguel (1997) writes,

> In seventeenth- and eighteenth-century Europe, it had been assumed that books were meant to be read indoors, within the secluding walls of a private or public library. Now, publishers were producing books meant to be taken out into the open, books made specifically for travel.
>
> (p. 141)

The shape of the book changed in part because of the increased mobility of the leisured bourgeoisie in Victorian England. Nineteenth-century England was in the throngs of industrialization, and one of the major forces of industrialization was the growth of the railway. The railway facilitated the transportation of goods and people and quickly supplanted the horse-drawn coach as the dominant form of transportation. Contemporary sources proclaimed that the railway "annihilated space and time" by compressing the travel time between two points, and the bourgeoisie took advantage of this compression by traveling more frequently, and further, than ever before (Schivelbusch, 1986), which led to a need to develop new ways to fill the increased travel time and control the social situation of the railway. The shape of the book changed to meet those needs, leading to the modern form of the paperback novel (Manguel, 1997).

The social situation of the railway was much different than earlier horse-drawn forms of transportation. On the railway, the bourgeoisie were faced with circumstances similar to the growth of the city outlined by Simmel: For the first time, people were expected to sit for long periods in the company of strangers. Before the railway, travel was done by carriage and most people knew each other in the carriage. The European railway changed that, in great part through the design of first-class railway compartments. Railway compartments comprised divided rooms that sat up to eight people in a U-shape. Each compartment contained two rows of seats that faced each other, meaning people had to sit across from complete strangers for long periods of time. According to Schivelbusch (1986), for many bourgeoisie travelers, the forced socialization with strangers was almost unbearable. Just as Simmel's metropolitan man developed the *blasé* attitude to deal with the chaos of the city, Victorian bourgeoisie turned to any type of interface that could give them control over their experience of the railway compartment. These interfaces were the newspaper and the portable paperback novel.

Books and newspapers allowed individuals to partially control their experience of the railway car by imparting the narrative of what they were reading on their perception of the shared space. They could avoid dealing with the other people in the compartment by using the book or newspaper as an object of subordinate involvement, allowing them to appear occupied and unavailable for conversation. According to Goffman (1963),

> This is one reason why waiting rooms, club cars and passenger airplanes in our society often are supplied by management with emergency supplies such as magazines and newspapers, which serve as minimal involvements that can be given weight (when there is nothing but waiting to do) yet can be immediately discarded when one's turn or destination arrives.
>
> (p. 51)

In the case of nineteenth-century trains (and, for that matter, all public transportation), the dominant involvement is the final destination the passenger wants to reach. But during the journey, the passenger might sustain "quite absorbing main involvements [with books and newspapers] which are patently subordinated to the dominant involvement" (Goffman, 1963, p. 51). As Goffman points out, the more one context's dominant involvement is obviously outside the situation (as is the case of riding in trains and airplanes), the more freedom the passenger has to engage in activities of subordinate involvement (such as reading books and newspapers) without being perceived as withdrawing from that which he or she should be involved in. Using book and newspapers as objects of subordinate involvement gave people a way to avoid visual connections with other passengers, and therefore manage their involvement with the present situation. They could use the book to filter out their surrounding space by focusing on the narrative of the paperback novel.

It would be easy to slip into using the idea of withdrawal to describe what one does when reading an engrossing novel on a train. We can likely all understand that line of reasoning—exemplified by phrases such as "I was lost in the novel." But the reader does not completely withdraw from the space of the train into the narrative of the novel. She is *both* in the train compartment and in the space of the novel. The experience of the narrative is shaped by the place she is sitting, as much as the experience of the place is shaped by the narrative. By immersing themselves in novels, readers are able to experience a place differently by, for example, feeling that a subway trip is more enjoyable or by imagining a coffee shop as a personal reading room. The place of the reader can also influence the experience of reading. For example, reading while sitting next to people talking loudly on a bus can influence the reader's concentration on the narrative. Neither can overcome the other, making it difficult to conceive of someone metaphorically "withdrawing" from space. At most, the book can fully occupy only one sense: the visual.[4] Readers still experience the place they occupy through the haptic, aural, and olfactory senses. Rather, the book works as a filter by refocusing the visual sense, the sense both Simmel and Schivelbush argue is the most uncomfortable when sharing space with strangers.

Books and newspapers were consciously used to avoid interacting with other stimulation (including other people) present not only in the railway compartment, but also in urban public spaces. But using books to avoid people is not the same as withdrawing from a place. While some readers can be "lost in a novel" temporarily, they can always be jolted back to the realization they are in the subway car or in a main square when reading a book, whether that jolt comes from a loud sound or a tap on the shoulder. Even the most immersed readers must still be somewhat responsive to their surrounding space or else they cannot function in public. Readers are still in public and must be somewhat responsive to their environment even as they enjoy that sense of control enabled by the book. People may be co-present with random others in the train compartment, or on a street bench, just as Simmel's urbanites had to deal with random strangers in the chaos of the street, but they can exert control over randomness by using the book to impart a certain kind of order.

For 150 years after the first paperback novel was printed, books and newspapers remained the most effective mobile technologies for filtering experiences of public space. The efficacy of print media as interfaces, however, was somewhat limited by their visual nature. The visual has often been identified as the dominant sense in Modernist epistemologies (Sterne, 2003), and most urban theory has been no different (Bull, 2000). Possibly the most famous piece ever written about the importance of the gaze in the city is Baudelaire's discussion of the *flâneur*. The *flâneur* wanders the streets of the city, watching the crowds, soaking them in, but never becoming part of the crowd. He (and it was always a he) experiences the city by *looking* at the people on the streets. As Adriana de Souza e Silva and Larissa Hjorth (2009) suggest,

> The *flâneur* wandered and consumed the city with detracted gaze. He provided another type of lens through which to read and participate in the city, exploring new angles and avenues in the shifting vignette of early postindustrial urbanity. In this new picture of modernity of which the *flâneur* played both the role of the distanced critic and immersed spectator, vision was central.

> (pp. 606–607)

Baudelaire's *flâneur* shaped a century of urban theory, from Walter Benjamin's (1980 [1929]) writings in the beginning of the twentieth century to recent discussions of the "phoneur" in mobile phone studies (Luke, 2006). The "phoneur," rejecting the *flâneur*'s emphasis on the visual, epitomizes the importance of the haptic and the aural in our interactions with urban spaces and new technologies, such as mobile phones, iPods, and Walkmen. Henri Lefebvre (1991) also persuasively argued that the ocular centrism implicit in many modernist urban theories has limited analyses of the effects built spaces have on social interactions and perceptions of space. He writes that "all of social life becomes the mere decipherment of messages by the eyes, the mere reading of texts" (p. 286). Spaces are never just texts to be taken in through the act of looking. We experience space with all of our senses, and the book only works as an effective interface for one of them.

The problem with using print media as a way to filter information in public spaces is that the visual is often not the dominant perceptory sense when occupying shared spaces. Of all the senses, vision is the easiest to control. People can focus on something in the periphery, look down at their feet, or simply close their eyes, but they have less control over other senses. Perhaps the most difficult sense to control without the use of external interfaces is the auditory, which Michael Bull (2000) argues is often the dominant sense through which we experience the urban. While it is true that the growth of the modern city led to an overload of visual stimulation in the form of throngs of strangers and new types of public advertising and electric light, the sounds of the city increased as well.

The sounds of the city were more difficult to manage than the visual, at least until the development of auditory mobile technologies such as the Walkman. Bull argues that the Walkman, even more than books and newspapers, allowed individuals to control their experience of space because the Walkman allowed people to control the auditory nature of a place. Sitting on a crowded train, a newspaper or a book can do little to filter another co-present person talking to somebody else. People may be able to close their eyes to avoid the visual stimulation of a place, but they cannot close their ears. Mobile auditory technologies, however, help people control their auditory experience of places.

On June 1, 1979, Sony released the first Walkman in Japan; one year later the Walkman came to the United States. The Walkman represented a technological feat of Japanese miniaturization, condensing stereo equipment into a portable technology that could fit in the palm of a hand. Sony marketed the Walkman mainly to young audiences through an advertising campaign that emphasized the ability to maintain a mobile, active lifestyle while listening to music. Walkman advertisements from that time feature young adults rollerblading, riding the bus, and walking through crowded streets, all with headphones in their ears and happy expressions on their faces.[5] The thrust of Sony's initial campaign was that the new portable stereo technologies allowed young people increasing control over their experience of space. They no longer had to remain tied down by their music; they could take it with them wherever they needed to go.

With today's popularity of iPods and other mobile auditory technologies, it is easy to forget the major impact of the combination of music and mobility. William Gibson (1993), the novelist who coined the term "cyberspace," wrote,

> The Sony Walkman has done more to change human perception than any virtual reality gadget. I can't remember any technological experience since that was quite so wonderful as being able to take music and move it through landscapes and architecture.

> (n.p.)

Notice that in Gibson's quote, it is not the headphones that make the Walkman such an important technological development; rather, it is the combination of the individual nature of the headphones with the mobility of the subject. Headphones, after all, existed long before the Walkman brought them out onto city streets. Jonathan Sterne (2003) traces the development of headphones back to the 1816 invention of the stethoscope. The "headphones" of the stethoscope provided doctors with the first instance of "'in-head' acoustic imaging" (Stankievich, 2007, p. 55). The subjective interiority of headphones was then adopted for different listening practices, ranging from telegraph operators to people listening to sedentary music playing devices. The major shift with the Walkman was that the individualized soundscape in the head of the listener could be transported through physical spaces. Sony realized this, and heavily marketed the technology as providing a new freedom that combined individualized soundscapes and mobility.

As with books and newspapers, the Walkman was ultimately a technology of control (Hosokawa, 1984). Simmel's (1950) metropolitan man longed to summon "the utmost in uniqueness and particularization, in order to preserve his most personal core" (p. 422); the Walkman provided that uniqueness, letting users control the sounds of the space they moved through. Bull (2004a, 2004b) argues that "Walkman use can make the environment 'what it is' for users . . . This is achieved by users *repossessing* space as part of, or constitutive of, their desire" (Bull, 2000, p. 37, emphasis added). We do not control whom we walk by on the street or whom we sit next to on a train, but with the Walkman we can choose what we listen to. We control the experience of the sounds of the train compartment or the street corner more effectively than print media ever allowed.

The addition of an auditory layer to space introduced by the Walkman was not experienced by all users in the same way. In his extensive empirical examination of Walkman use, Bull (2000) shows the different ways in which people used the Walkman to alter their perception of space. He talks about how many people used the Walkman to soundtrack their movement through the city with their favorite music. For these users, the Walkman acted as an interesting aestheticizing force because users did not simply ignore what was around them, but instead changed the way they perceived what was around them. Walking down the street, the soundtrack laid over that experience changed the perception of the street. By filtering out the sounds of the city and replacing them with a personalized soundscape, the spaces of the city became almost filmic. Many users felt like they were walking through a movie where the other people on the streets were responding to the soundtrack being played through their headphones (Bull, 2000). So, as with the book, the Walkman did not lead to a "withdrawal" from public space; instead, it reshaped the experience of place, personalizing it and filtering out the auditory nature of the environment.

Once again, the difference between withdrawing from space and filtering space is important for how we understand mobile interfaces. Listening to music does not mean users no longer inhabit physical space. They do lose an auditory connection with the place in which they are, but that does not mean they are not present in that space. Blocking out external stimulation does not mean we do not experience places, it means we experience places *differently*. Walkman users remained conscious of the places they occupied, and just as the experience of a place is shaped by the personalized soundscape, so is the soundscape shaped by a place.

Both the Walkman and the iPod, which we discuss below, can be thought of as technological forms of the micro-behaviors Goffman analyzes in his discussions of behavior in public spaces. They are visible "do not disturb signs" expressing listeners' unwillingness to engage with others in shared space. Both technologies function similarly to what Goffman (1997) calls the conversational preserve: The "right of an individual to exert some control over who can summon him into talk and when he can be summoned" (p. 51). Putting headphones in one's ears is a clear social marker that conversation with others is not welcome. According to Sennett (1977), the right to be left alone in public is a relatively recent development tracing back only to the nineteenth century: "There grew up a notion that strangers had no right to speak to each other, that each man possessed as a public right an invisible shield, a right to be left alone" (p. 27). The increasing use of mobile auditory technologies in public is related to that development; headphones are a visible instantiation of the perceived right of urban individuals to move through public spaces while maintaining control over who can and cannot engage with them.

Thirty years after its release, the Walkman has moved from cultural icon to historical artifact. The dominance of the Walkman was supposed to wane as portable music began moving to digital formats with the CD player. CD players and cassette players, however, co-existed for two decades until the move in the early 2000s to MP3 players. The first MP3 player was the Listen Up Player, introduced by Audio Highway in 1996. But the MP3 format did not achieve dominance until the release of the Apple iPod in 2001, which became a cultural icon of the 2000s (Bull, 2007). The iPod became such a dominant form of mobile media for many reasons, one of the most notable being the popularity of the file-sharing site Napster. The original idea behind the iPod was to "take an MP3 player, build a Napster music sale service to complement it, and build a company around it" (Kahney, 2004 , n.p.). With the growth of MP3 distribution channels, iPod users had access to more music than was practically possible with cassettes and CDs. The storage capabilities of the devices and the development of Napster, followed by iTunes, allowed for an increased freedom for listeners. With the iPod, vast music libraries could be stored, and users could set up playlists and use the "shuffle" option so that songs did not have a pre-defined order anymore. Therefore, listeners no longer had to decide what they wanted to listen to before they left the house because they could carry their entire music library with them. Of course, it was not just storage space and the ability to shuffle songs that made the iPod so successful. The iPod also became such an icon because of the cultural cache Apple had built over the years and the functionality and aesthetics of the device (Cornwall & Gaventa, 2001).

While it may seem that the move from the Walkman and CD players to the MP3 format of the iPod was only a minor shift, understanding the iPod as only a small improvement over the Walkman is a mistake. The iPod's affordances mark a qualitative shift in how individuals use technologies to interact with space. People still put on headphones, turn their music up, and soundtrack their space, but iPods give users greater control over how they interact with that space. While the Walkman lets people put a tape in and play it in public space, iPods—because of the large amount of music they hold and their processing power—let people customize their music choice to match the spaces they move through. In other words, a Walkman user had to choose which tape to bring before she left her house; an iPod user carries her "auditory identity in the palm of her hand" (Bull, 2006). Instead of overlaying the space with music they had to choose before leaving (a cassette), iPod users can draw from their music libraries while mobile and match their music to the space they move through, giving them increased power to construct how they experience public space. As Bull (2007) puts it,

> [iPods] permit a transformation and control of the user's everyday experience. The Apple iPod does this more successfully than more traditional mobile devices as the user synchronises the world to their own private soundworld—the world walks in step to the Ipod user.
>
> (p. 137)

The iPod, just like the Walkman, has also been derided by cultural critics. For example, Armstrong Williams (2006), writing for the *New York Amsterdam*, claims that using mobile technologies such as the iPod while in public is "unhealthy, unproductive, and downright rude." He then goes on to say,

> When you plug into your iPod in a public place, you are basically telling everyone else that you do not want to interact with them. You are telling the world that you want to be isolated and left alone.
>
> (n.p)

But just like with the Walkman, critics such as Williams imagine the iPod destroying a much healthier public life than actually exists. As the previous two sections show, the iPod does not exist in a cause and effect relationship to declining urban sociability. Attributing the decline of public life to the iPod and the Walkman would be too strong a claim. Rather, the built environment—exemplified by growth of megalopolises, privatized infrastructure, isolated urban enclaves, and the valuation of individualism—have all led to the decrease of random sociability. As Erkki Huhtahmo (2004) argues, "The urban experience had to go through many changes before auditory seclusion from the surroundings became an accepted and desired practice" (p. 4). It is incorrect to argue that technologically mediated experiences in public spaces directly lead to their decline, or that public spaces are void of sociability. As we will see later in the book, mobile interfaces also promote different types of social interactions with people and locations, leading to the construction of new types of spaces filled with meaning and sociability.

Criticisms of mobile media are not all misplaced. Technologies such as the Walkman and the iPod do likely have some atomizing effect on urban individuals. Headphones allow people to avoid having to deal with others, and one can form their own opinion on the seriousness of that avoidance. As we have pointed out in this chapter, however, many of these criticisms fall back on a nostalgia for a public life of the city that disappeared long before mobile auditory media became ubiquitous. Another problem with some of the criticisms of mobile technologies is the way in which they argue that the technologies adulterate how individuals experience space. As we argue throughout this book, all experiences of space are mediated, and mobile technologies are best viewed as interfaces to these spaces, allowing different perceptions of places, and locations.

A Filtered World

We all have our idealized places, the places to which we feel the strongest connection. Whether that place is a mountaintop overlooking a verdant valley or a crowded public square permeated by the mist of a central fountain, some places do grasp our attention better than others. Most places we move through, however, fall far short of this ideal. They do not demand our full attention, nor would devoting ourselves fully to them be a particularly pleasurable experience. We develop techniques to interface our interaction with these places, ways to divert our attention. There are many ways to do this, including Simmel's *blasé* attitude, Sennett's turn toward a controlled, private experience, or Goffman's structure of micro-behaviors in public, as well as technologically mediated techniques we use to alter the way in which we engage with the places we move through.

While some critics argue that mobile technologies lead to a disconnection from physical space, there is an equally strong counter argument by which we consider mobile technologies as an intrinsic *part of* people's experience of space, and as we will see in later chapters, new location-aware mobile technologies work through their interaction with spatial information. Discussing mobile technology use in public space, Paul Goldberg (2007) writes, "you are not on Madison Avenue if you are holding a little object to your ear that pulls you toward a person in Omaha" (n.p). Most people who listen to music while moving through the streets do so in a conscious effort to ignore the sounds of the streets. They are conscious of their physical surrounding, and that consciousness is why they use mobile technologies in the first place. Undoubtedly, reading a book or listening to music can be distracting. We lose one of our sensorial connections to places when we read or listen to auditory media in public, but we also become distracted when we walk down the street while in a serious conversation with the person next to us. Either way, we do not completely lose our sense of connection to the surrounding space. We cannot only be in Omaha when we are walking in Manhattan.

The larger issue here concerns the meaning of interface. The experience of a place interfaced through mobile technologies has been criticized as being less pure. The same arguments are present when it comes to communication, where different forms of computer-mediated communication are measured against face-to-face communication, the most "pure" form of communication (Peters, 1999).

But the issue is not one of a pure, unadulterated connection to space versus a withdrawn, mediated experience of space. Every experience of space is a mediated experience. As Hansen (2006) points out, human skin was the original medium, or in our terms, the original interface. Everything we know about the spaces we inhabit, every sensation we take in, is mediated by our bodies. The mobile technologies described here only add a layer of mediation to our experience of space. It is true that each added layer changes the meaning of our interactions both with the space that surrounds us as well as with other people who inhabit these spaces, but these technologies do not make the sense of space somehow impure, and they do not withdraw us from the surrounding space. The socio-technical situation does blur the lines between spaces, allowing people to exist in the space of the song and the corporeal space of the street, but it is key to remember that it is *both at once*, not one or the other.

The connection between technology use, physical location, and experience of space is a key point driving this book. The connection becomes more obvious when we move on to address location-aware mobile technologies in the forthcoming chapters, but it is important to remember that people still use older mobile technologies as well to reconfigure their relationship to the spaces they inhabit and move through, not negate the importance of them. Wearing headphones while walking through the streets may change the street from a place of random sociability into a place made to fit the soundscapes we choose, but we still exist through our bodies, and our bodies must always be "in place" (Casey, 2009).

Notes

1   Specifically in Chapter 3, we address how mobile phones have been often viewed as disconnecting users from their surrounding space.

2   Mixed-use development means a neighborhood or a section of a city is used for more than one purpose. For example, a mixed-use area could include shops, apartments, and offices. Mixed use stands in contrast to the typical way of building suburbs, in which housing is segregated from business areas.

3   As we will see in Chapter 3, many mobile communication scholars also studied the mobile phone as an instrument that helps the development of strong ties, and consequently leads to the neglect of other weak ties in public spaces (Hampton & Gupta, 2008; Hampton, Livio, & Sessions, 2009; Ling, 2008; Wellman, Quan Haase, Witte, & Hampton, 2001).

4   There are those who claim that there is a kind of synesthesic immersion, which is possible while reading good literature (e.g., the idea that Dickens gets his readers smelling the smells and hearing the sounds of London) or, at least, that one can become so immersed that she is able to "switch off" the senses from the physical environment (as when someone engrossed in a novel literally does not hear someone speaking to them). However, this is a kind of mental state (which we argue is comparable to the *blasé* attitude) that can be switched on and off by the reader. Want it or not, the reader still experiences the place she occupies.

In literary theory, there is a concept of *immersion* that argues that people can be immersed in a "fictional world which is largely the product of our own mental, cognitive, abilities to create that fictive, virtual (in the figurative sense of the word) world from the symbolic representations—the text, whether purely linguistic or multi-modal, digital or print—displayed by means of any technological platform" (Mangen, 2008, pp. 404–419). We do not argue that a great novel or a great song cannot immerse someone in their own mental world. However, we view immersion as a way in which individuals use media to help manage some of the chaos of their surrounding space by refocusing their attention on the fictive worlds in which they are immersed.

5   For a sample of Sony's early Walkman advertisements, see the following sites: http://www.grayflannelsuit.net/retrotisements-rip-sony-walkman-1979-2010/ and http://gizmodo.com/5305177/great-sony-walkman-tv-and-print-ads-of-the-1980s.

References

Baudelaire, C. (1964). *The painter of modern life*. New York: Da Capo Press.

Benjamin, W. (1980 [1929]). Die Wiederkehr des Flaneurs. *ders.: Gesammelte Schriften, Bd. HI, Frankfurt a. M. (Suhrkamp Verlag)*.

Bouwhuis, J. (2007). Nettime: Interview with Norman Klein on the New Canon. October 2. Retrieved March 7, 2011from http://www.16bcavergroup.org/mtarchive/archives/002324.php.

Bull, M. (2000). *Sounding out the city: Personal stereos and the management of everyday life*. Oxford: Berg.

Bull, M. (2004a). Sound connections: An aural epistemology of proximity and distance in urban culture. *Environment and Planning D, 22*, 103–116.

Bull, M. (2004b). Thinking about sound, proximity, and distance in Western experience: The case of Odysseus's Walkman. In V. Erlmann (Ed.), *Hearing cultures: Essays on sound, listening, and modernity*. New York: Berg.

Bull, M. (2007). *Sound moves: iPod culture and urban experience*. New York: Routledge.

Canclini, N. G. (2001). *Consumers and citizens: Globalization and multicultural conflicts*. Minneapolis: University of Minnesota Press.

Casey, E. (2009). *Getting back into place: Toward a renewed understanding of the place-world* (2nd ed.). Bloomington, IN: Indiana University Press.

Cornwall, A., & Gaventa, J. (2001). From users and choosers to makers and shapers: Repositioning participation in social policy. Paper presented at the Institute of Development Studies.

Crary, J. (2001). *Suspensions of perception: Attention, spectacle, and modern culture*. Cambridge, MA: The MIT Press.

De Certeau, M. (1988). *The practice of everyday life*. Minnesota: University of Minnesota Press.

de Souza e Silva, A., & Hjorth, L. (2009). Playful urban spaces: A historical approach to mobile games. *Simulation and Gaming, 40*(5), 602–625.

Fischer, C. S. (2005). Bowling alone: What's the score? *Social Networks, 27*(2), 155–167.

Gergen, K. (2002). The challenge of absent presence. In J. Katz, & M. Aakhus (Eds.), *Perpetual contact: Mobile communication, private talk, public performance* (pp. 227–241). New York: Cambridge University Press.

Gibson, W. (1993). Time Out. *Time Out*, 49.

Goffman, E. (1963). *Behavior in public places: Notes on the social organization of gatherings*. New York: Free Press of Glencoe.

Goffman, E. (1997). The Goffman reader. In C. Lembert, & A. Branaman (Eds.), *The Goffman reader*. Malden, MA: Blackwell.

Goldberg, P. (2007). Disconnected urbanism. *MetropolisMag.com*. Retrieved from http://www.metropolismag.com/story/20070222/disconnected-urbanism.

Goldhaber, M. (2006). The value of openness in an attention economy. *First Monday, 11*(6).

Gordon, E., & de Souza e Silva, A. (2011). *Network locality: How digital networks create a culture of location*. Boston: Blackwell Publishers.

Graham, S., & Marvin, S. (2001). *Splintering urbanism: Networked infrastructures, technological mobilities, and the urban condition*. New York: Routledge.

Hampton, K., & Gupta, N. (2008). Community and social interaction in the wireless city: Wi-fi use in public and semi-public spaces. *New Media and Society, 10*(8), 831.

Hampton, K., Livio, O., & Sessions, L. (2010). The social life of wireless urban spaces: Internet use, social networks, and the public realm. *Journal of Communication, 60,* 701–722.

Hansen, M. (2006). *Bodies in code: Interfaces with digital media*. New York: Routledge.

Hobart, M., & Schiffman, Z. (1998). *Information ages: Literacy, numeracy, and the computer revolution*. Baltimore, MA: Johns Hopkins University Press.

Hosokawa, S. (1984). The Walkman effect. *Popular Music, 4,* 165–180.

Huhtamo, E. (2004). An archeaology of mobile media. Paper presented at the Keynote address at ISEA.

Humphreys, L. (2007). Mobile social networks and social practice: A case study of Dodgeball. *Journal of Computer-Mediated Communication, 13*(1), article 17.

Jacobs, J. (1961). *The death of life of great American cities*. New York: Random House.

Kahney, L. (2004). Insider look at birth of the iPod. *Wired*. Retrieved from http://www.wired.com/gadgets/mac/news/2004/07/64286.

Lanham, R. (1997). A computer-based Harvard red book: General education in the digital age. In L. Dowler (Ed.), *Gateways to knowledge: The role of academic libraries in teaching, learning and research*. Cambridge, MA: MIT Press.

Lanham, R. (2006). *The economics of attention: Style and substance in the age of information*. Chicago: Chicago University Press.

Lefebvre, H. (1991). *The production of space*. Malden, MA: Blackwell Publishers.

Ling, R. (2008). *New tech new ties: How mobile communication is reshaping social cohesion*. Cambridge, MA: MIT Press.

Luke, R. (2006). The phoneur: Mobile commerce and the digital pedagogies of the wireless web. In P. Trifonas (Ed.), *Communities of difference: Culture, language, technology* (pp. 185–204). London: Palgrave Macmillan.

Mangen, A. (2008). Hypertext fiction reading: haptics and immersion. *Journal of Research in Reading, 31*(4), 404–419.

Manguel, A. (1997). *A history of reading*. New York: Penguin Books.

Packer, J. (2008). *Mobility without mayhem: Safety, cars, and citizenship*. Durham, NC: Duke University Press.

Peters, J. D. (1999). *Speaking into the air: A history of the idea of communication*. Chicago: University of Chicago Press.

Plant, S. (2001). *On the mobile: The effects of mobile telephones on social and individual life*. London: Motorola Inc.

Puro, J. P. (2002). Finland, a mobile culture. In J. Katz, & M. Aakhus (Eds.), *Perpetual contact: Mobile communication, private talk, public performance* (pp. 19–29). Cambridge: Cambridge University Press.

Putnam, R. D. (2000). *Bowling alone: The collapse and revival of American community*. New York: Simon & Schuster.

Schectman, J. (2009). Iran's twitter revolution? Maybe not yet. *Business Week*, June 17.

Schivelbusch, W. (1986). *The railway journey: The industrialization of time and space in the 19th century*. Berkeley, CA: University of California Press.

Scott, A. J., & Soja, E. W. (1998). *The city: Los Angeles and urban theory at the end of the twentieth century*. Los Angeles: University of California Press.

Sennett, R. (1977). *The fall of public man*. New York: Knopf.

Sennett, R. (1992). *The conscience of the eye: The design and social life of cities*. New York: Norton.

Simmel, G. (1950). *The sociology of Georg Simmel* (K. Wolff, Trans.). New York: Free Press.

Soja, E. W. (1996). *Thirdspace: Journeys to Los Angeles and other real-and-imagined places*. Boston: Blackwell.

Stankievich, C. (2007). From stethoscopes to headphones: An acoustic spatialization of subjectivity. *Leonardo Music Journal, 17*, 55–59.

Sterne, J. (2003). *The audible past: Cultural origins of sound reproduction*. Durham, NC: Duke University Press.

Urry, J. (2000). *Sociology beyond societies: Mobilities for the twenty-first century*. New York: Routledge.

Urry, J. (2007). *Mobilities*. Cambridge: Polity Press.

Weber, M. (2003). *The Protestant ethic and the spirit of capitalism* (T. Parsons, Trans.). Mineola, NY: Dover Publication.

Wellman, B. (2002). Little boxes, globalization, and networked individualism. In M. Tanabe, P. Van den Besselaar, & T. Ishida (Eds.), *Digital cities II: Computational and sociological approaches* (pp. 10–26). Berlin: Springer.

Wellman, B., Quan Haase, A., Witte, J., & Hampton, K. (2001). Does the internet increase, decrease, or supplement social capital? Social networks, participation, and community commitment. *American Behavioral Scientist, 45*(3), 436.

Whyte, W. (1980). *The social life of small urban spaces* Washington, DC: Conservation Foundation.

Williams, A. (2006). Americans experiencing technological overload. *New York Amsterdam*. Retrieved from http://archive.newsmax.com/archives/articles/2006/5/18/144041.shtml.

Wirth, L. (1938). Urbanism as a way of life. *American Journal of Sociology, 44*, 1–24.

2

THE PUBLIC AND THE PRIVATE

Johanna leaves the baggage claim area at the airport and looks for the subway. It is Saturday and the airport is more crowded than usual. She never liked crowds, but a crowded place in an unknown city makes her even more anxious. The signs to public transportation are not clear, and because she is in a new country, turning on the data features of her iPhone to look for directions is expensive. She has no choice but to ask some random stranger for directions.

She finally finds the train and takes it toward downtown. After a long flight across time zones, she is extremely tired. All she wants to do now is to find the apartment she had rented online, eat some food, and sleep. She wants nothing more than for the train ride to be finished. A woman comes in, and sits by her side. She has lots of bags and occupies two-thirds of the seat, making Johanna slide closer to the window. Another group of about ten teenagers enters the train, probably going to some party or bar downtown. They are speaking loud and laughing, and they don't sit down. Almost immediately after the young people arrived, a man in his thirties sits across the aisle. He turns on some random pop-music on his phone—without headphones. "Maybe they are competing to see who can be louder," thinks Johanna. Another man complains about the noise because he is trying to read. They ignore him.

She then decides to listen to her iPod, thinking that maybe she can ignore everything else until she gets to her station. After listening to music for a couple minutes, she begins to feel better and is relieved that she no longer has to listen to the random sounds of the train. She entertains herself by watching the teenagers interact with each other, but without really knowing what they are talking about. She tries to imagine a fictitious dialogue between a boy and girl who seem to be flirting with each other and almost forgets how tired she is. Faster than she expected, the train arrives at the downtown station. She takes one headphone out of her ear, and gets off the train. Her new apartment is two blocks from the subway station.

Mobile technology use in public spaces has often been perceived as an invasion of the public by the private, and the divide between public and private spaces is a problematic one. Norbert Bobbio (1989) labels this divide one of the "grand dichotomies" used to understand the social world. As with all dichotomies, the separation of two terms on opposite poles obfuscates a wide range of distinctions and variance. In between two pieces in a binary opposition there lies much fruitful ground that is ignored when objects or situations are forced into one of two boxes.

Theorists who use the public/private dichotomy face a more serious problem than the broad strokes necessary for any theoretical dichotomy. Namely, public and private mean very different things to different people. For example, for many urban theorists public spaces are shared open spaces, such as plazas and squares. For many social theorists, they are spaces for public deliberation, following the model of the Ancient Greek Agora. In the eighteenth century, these public spaces could be cafes, where much of the political talk took place, but for economists, cafes are private spaces because they are owned by private individuals or corporations. However, for feminist theorists, private spaces equate to the secluded space of the home. Regardless of which perspective one adopts, public and private are always socially constructed, and thus shift across cultures and time. What is considered private and what is considered public has changed through history, and the borders between the two have always been permeable.

The private and public divide is also challenged as new technologies are introduced into the social fabric. Mobile technology use in public spaces complicates traditional understandings of what it means to be in public, allowing people to bring previously private activities (reading, listening to music) into public spaces. When people sit on a crowded train with headphones in their ears, is the space as public for them as it is for the two people sitting two rows behind who are engaged in a conversation? Or does the space become private, personal, and controlled? The answers to these questions depend on how we understand the terms "private" and "public."

In this chapter, we build on the previous chapter's examination of the relationship between mobile media and space by discussing the use of mobile technologies in public spaces and situating our approach to public spaces within a discussion of the public/private dichotomy. So far we have examined how mobile technologies work as interfaces to public spaces by mediating people's relationship to these spaces and to other people in them. Mobile technologies can be viewed as interfaces because their use also influences the meanings and perceptions of public spaces. We have been using the term "public space" quite a lot, but we have not discussed what we mean by public space. In this chapter, we define our approach to public spaces by contrasting it to the idea of private spaces. We show how the idea of public and private has assumed different meanings throughout history. For example, Jeff Weintraub (1997) characterizes four different models through which the public/private dichotomy has been conceptualized: the citizenship model, the feminist model, the economic model, and the sociability model. We re-frame this multifaceted approach to discuss how mobile technologies—namely, the portable camera, the book, the Walkman, and the iPod—complexify commonly understood divisions between private and public experience.

In the first chapter, we discussed the book, the Walkman, and the iPod as mobile interfaces individuals use to control their interactions with public spaces. In this chapter, we analyze how mobile interfaces complicate the very notion of public space. We do so by examining different theorists who have discussed how mobile technologies challenge the public/private divide, most often arguing that mobile technologies privatize public spaces (Bull, 2000, 2007; du Gay, Hall, Janes, Mackay, & Negus, 1997). But we disagree that what is occurring is only a privatization of public space. Certainly, mobile technology use in public spaces may help people feel more familiar with those spaces. But, more importantly, mobile technologies make the socially negotiated nature of public and private visible. They do not lead to the death of public space. Instead, mobile technology use is a physical instantiation of the constantly negotiated understandings of how public and private are related. Mobile technology use in public spaces foreground the already blurred lines between private and public, forming what Mimi Sheller and John Urry (2003) call "private-in-public" hybrids. As Bruce Bimber, Andrew Flanagin and Cynthia Stohl (2005) suggest, "as individuals are able to move more seamlessly between private and public domains, the structure of public domain themselves is altered" (p. 382). We therefore end the chapter by discussing the role of mobile technologies in (re)constructing the meaning of public spaces.

The Different Understandings of the Private and the Public

Public spaces are shared spaces. They exist outside the domestic sphere. We understand public spaces as spaces where strangers congregate and where heterogeneous individuals co-exist. Following Jane Jacobs (1961), we view public spaces as social spaces, comprising "fluid sociability among strangers and near strangers" (p. 17). Anthony Giddens (1991) discusses how Richard Sennett (1977) has argued that "the words 'public' and 'private' are both creations of the modern period" (Giddens, 1991, p. 151). Giddens points out that Sennett may be correct in one sense, but his argument misses the different ways in which the private and public have been used as organizing logics of social, political, and economic life throughout history.

The tension between the public and the private traces back much further than Sennett suggests. In Ancient Greece, Aristotle (2010) saw this tension as the split between the household (*oikos*) and political life (the *polis*). Similarly, Hannah Arendt (1958) discusses the Ancient Greek divide between the public realm and private life. In Greek society, the public was the place for politics, such as the Agora, where things "could be seen and heard by everybody" (Arendt, 1958, p. 58). The private, however, was the place of property and the family. Private spaces were then secluded closed spaces, separate from the public open spaces.

Arendt's discussion of the Ancient Greek *polis* and its creation of public and private spaces is closely aligned with what Weintraub (1997) calls the citizenship model. For thinkers such as Arendt, Aristotle, and Habermas, the "public" was the setting for democratic deliberation and discussion. Public space in this model is viewed as distinctly political. Arendt labeled it the "public realm" and Habermas (1989) called it the "public sphere," both of which exist in relation to the non-political (at least in a direct sense) nature of private domestic life. Arendt's public realm was, to a degree, physically bounded: She identified the Greek Agora as the site of public life. For Habermas, the public sphere was not strongly dependent on a bounded physical space. The prototypical public sphere for Habermas was the cafe life of the eighteenth century, but Habermas also viewed the political discourse that took place in eighteenth-century print culture as a "site" for the public sphere. Internet theorists have also argued that websites can function as vital pieces of the public sphere, showing that public spaces do not necessarily take place in physically bounded spaces (see Burrows & Ellison, 2004).

Arendt's and Aristotle's linking of the private with the domestic also has similarities with what Weintraub identified as the feminist model, which emerged around the 1970s most notably in the work of Michelle Zimbalist Rosaldo (1974). In the feminist model, private is most commonly associated with the family and domestic life to the point that Weintraub argues that domestic and private are almost interchangeable. Public, consequently, becomes everything outside the domestic sphere. One of the problems that feminist critics saw with the private/public split is that the public tends to be valorized over the private. The private is the space society has traditionally associated with women and the public becomes the sphere in which males operate; however, unlike with Weintraub's citizenship model, the public in the feminist model encompasses most everything (especially economic life) outside the household, not just political life. This is in direct contrast with the economic model of thinking about the distinction between public and private (Weintraub, 1997). In the feminist model, the private is the domestic sphere. But in the economic model, private belongs to the market forces of businesses and individual actors, while the public is the realm of governmental programs and services. In the economic model of thinking, the public typically represents the state, as in "public sector" or "public goods," and the private refers to the market, as can be seen in the oft used term "private sector."

While Weintraub does not directly mention "public space" in his discussion of the economic model, some ideas of public versus private spaces do partially fit inside this model. For example, critics have argued that public space is increasingly being replaced by private spaces, such as malls and other shopping centers (Davis, 2006; Freeman, 2008). In these arguments, public space is equated with spaces run by the government, whether at a local or federal level, while private spaces are spaces owned by private actors. We, however, do not use the term "public space" to refer only to spaces owned and controlled by the public sector. Instead, we think of public spaces as anywhere strangers often congregate. This is more in line with what Weintraub called the sociability model, which is closely tied to the work of Richard Sennett.

In the previous chapter we discussed Sennett's (1977) argument that the private life of intimate ties had encroached upon the public life of street sociability. Sennett, along with Philippe Aries, Georges Duby (Aries and Duby, 1987) and others, is one of the strongest proponents of the understanding of the public as social spaces. Jane Jacobs (1961) can also be put into this camp with her discussions of the public street life of American cities. Public life for Sennett and Jacobs is not represented by the collective action of strangers, as with the citizenship model; it is instead the co-existence of heterogenous individuals. Public life, then, is in direct contrast to the private domestic life, and the tight-knit community. The sociability model is also tied closely to issues of physical space. Jacobs wrote extensively on ways in which urban planning could encourage a healthy urban life, and her influence was later taken up by William Whyte (1980) and others. The understanding of public space as the world of random sociability, and private space as more intimate and controlled setting, is the way we use the terms throughout this book.

Public space, then, is often understood as the space where strangers co-exist, a setting over which individuals have little control. Private space thus becomes a controlled space,[1] a space where interactions with others tend to happen on terms that are more comfortable to the individual. However, public spaces are not only physical spaces. For example, a virtual world or a chat room can be considered a public space. In fact, danah boyd reframed the sociability model so that it also addresses online sociability, or "networked publics," as she names it. In her formulation of the concept of "networked publics," boyd (2007) argues that social networking sites are now the site where many teens learn how "to present themselves, and take risks that will help them to assess the boundaries of the social world" (p. 137). But in either case (a street or an online space), public spaces are sites of heterogeneity and co-existence of strangers. Also as it will become clear throughout this chapter, the borders between public and private spaces, although so often presented as a sharp divide, are in fact much more fluid and complex. What is considered private and what is considered public changes with time period, cultures, and the interfaces we use to interact with these spaces. The use of mobile technologies challenges the traditional borders between public and private spaces because individuals are able to interface with their experience of space in new ways, as they co-exist with others in public (and private) spaces.

Mobile Technologies, Public Spaces, and Privacy

Instead of looking at the tension between the public and the private as a divide or a dichotomy, it is more productive to the study of mobile technologies to understand the shifting boundaries between the two as fluid and permeable. It may be more accurate, as we discuss below, to say that many situations involve hybrid forms of public and private that throw into question the analytical accuracy of the dichotomy. Private and public also are often subjective states of mind rather than any actually demarcated spaces, something that Weintraub's four-part model does not discuss. Marking a space as public does not mean people will not have private experiences in that space, such as a couple carrying on a secret affair in the middle of a crowded restaurant. Likewise, just because a space such as a hotel room is supposedly private does not mean that people will automatically feel they possess the types of privacy typically associated with private spaces. However, despite the slipperiness of the terms, Allen Wolfe (1997) still urges that they are necessary for understanding the organization and experience of daily life. They have long been organizing logics we have used to understand the shape of our social worlds (Bobbio, 1989). Understanding the permeable nature of these concepts may be more productive than abandoning the terms altogether.

Take the example of the home, which is the prototypical private space in Western society. The home is where people can retreat from public life into the safety of their intimate relationships. While feminist critics have made great strides in moving past the idealization of home, it is still often considered the most private of spaces. The idea of home as private, however, is a relatively recent construction, one closely linked to technological developments. As Stuart Shapiro (1998) explains, it was not until the seventeenth century that people could really enjoy privacy in their homes. In medieval times, homes were not understood as bounded places that could only be entered by certain residents and contemporary understandings of privacy were more or less non-existent. On wedding nights, for example, the bride and groom would lie naked in bed in front of members of the community and consummate their marriage.

The way we understand privacy has greatly changed since those interesting medieval wedding nights, a change that can be closely linked to both new architectural layouts and the rise of individualism in the Modern period. In the sixteenth century, the "physical structure of the home reflected the low priority attached to individual privacy, as dwellings often consisted of a single communal room and were almost always crowded" (Castells et al., 2007, p. 278). At the beginning of the seventeenth century, more houses started being built with separate rooms inside the dwelling, allowing for increased privacy, but these rooms were still accorded a level of privacy much lower than what we are now accustomed to because houses were still rarely built with hallways. To cross through the house, one had to move through each room. It was the addition of the hallway that really shaped the house as a set of private spaces in the way we understand it today because people could move around the house without needing to walk through all the rooms. The hallway enabled people to enjoy what was closer to a more contemporary experience of the private spaces of the home.

The construction of hallways highlights the permeability of private and public, even inside what is often considered the exemplar of a private space. Note that hallways did not *cause* new forms of privacy. There had to be at least an incipient desire for that privacy for people to begin planning hallways (Shapiro, 1998). Nonetheless, examining the interrelationships between technology use (in this case, the house) and the shifting boundaries between private and public is important because often technology use is closely related to a shift in people's perceptions and understandings of public and private spaces. Communication technologies are also strongly related to these boundaries as well. Take the example of the telephone. The telephone allowed individuals to call into the space of the home, an act that was perceived as blurring the boundaries between public "outside" spaces and the private spaces of the home (Marvin, 1988). Broadcast media worked in a similar way. Through radio, and later television, the public could be brought into the private space of the house, further evidencing the permeability of the boundaries between the two concepts (Foucault, 1999).

The television complicates these boundaries by contributing to what Raymond Williams (1975) calls "mobile privatization." Television allowed people to remain in the private spaces of the home even as they watched the more public life of broadcast media from the comfort of their couches. Before televisions and radios, a shared experience of a stage play or a musical performance had to be experienced in a shared space with other people. With the television, people could "travel" to the performance and watch it as others were watching it without ever leaving their homes. The television contributed to the paradox that individuals were increasingly living both a more private and a more mobile life, traveling to far off performances through the television without having to leave their living rooms. Mobile privatization made the public more private by letting people experience public performance from the privacy of their homes. But it also made the house less private and, like with the telephone and the radio, allowed outside entities to reach into the home. These communication technologies did not turn domestic spaces into public spaces, but they did signal a shift to a new hybridized form of public-in-private that challenged the traditional public/private borders. Mobile technologies invert this situation. They take private activities into public spaces. At the simplest level, they do so because they are mobile. That may not say much we do not already know, but it is the mobility of these technologies that places them in an interesting relationship with accepted understandings of the private and the public. When technologies become mobile, they often leave the confines of places commonly understood as more private (e.g., the house, the office) and can be taken into public spaces, spaces shared "among strangers and near strangers" (Weintraub, 1997, p. 17). However, while mobile technologies bring private activities into public spaces, they can also make the public *even more public* by challenging pre-established notions of privacy. We can see this through an examination of an early type of mobile technology: the portable camera.

The Portable Camera and New Publicity

With the development of celluloid roll film and more portable, usable box cameras, the late 1880s witnessed the growth of amateur photography. George Eastman used his considerable business skills to market his new Kodak camera, and soon people were wandering the streets taking snap shots of whatever passed by (Figure 2.1). As media archeologist Erkki Huhtamo (2004) argues, these new amateur photographers often challenged social expectations through their camera practices, leading to what contemporary critics called "The Camera Epidemic."

The portable camera, like the mobile technologies we discussed earlier, basically untethered previously fixed practices. Cameras had existed long before Eastman and others began developing easy to use, portable cameras. Earlier cameras, however, had to be operated by professionals and were usually too bulky to be easily moved and taken out in public. Consequently, photography was a practice that typically took place in more closed, private spaces. The portable camera took photography out into the streets, allowing people to complicate societal norms by taking pictures of unsuspecting individuals. A robust market even arose for candid, hidden cameras "imaginatively disguised as hats, walking sticks, bags, and—yes—pocket watches" (Huhtamo, 2004, p. 3).

The simple description of how amateur photography disrupted societal norms would be to say that it violated individuals' right to privacy. In a way, it certainly did. But in another way, we can look at the portable camera as playing a major role in establishing that right to privacy. Before the portable camera, there was no explicit writing in U.S. legal scholarship about the rights individuals had regarding privacy. Privacy had certainly been discussed long before the portable camera became popular in the 1880s, but there was no expected right to privacy in the U.S. legal thinking (Solove, 2008). In 1890, that changed with Samuel Warren and Louis Brandeis' legal review article "The right to privacy," still one of the most important legal reviews in U.S. legal history.

Figure 2.1 Advertisement for the Kodak camera in the *Overland Monthly Magazine* from August 1889. Copyright: University of Michigan's Library Making of America.

"The right to privacy" begins with a rather teleological argument, tracing the development of rights and laws as if each new step was inevitable. Warren and Brandeis' (1890) brief history of law then arrives at their contemporary moment, and they write, "Recent inventions and business methods call attention to the next step which must be taken for the protection of the person, and for securing to the individual . . . the right to be let alone" (p. 195). The "recent inventions and business methods" were the development of the portable camera and more scandalous approaches of newspapers. Newspapers had begun to focus more and more on gossip, and they supplemented this gossip through photographs of unsuspecting people. No one was safe from the portable camera. As John Durham Peters (1999) writes, "privacy, quite explicitly, emerges as a concern once it is threatened by new media of image and sound recording" (pp. 174–175).

The ability to capture images while in public played a major role in the establishment of the right to privacy in the United States, but how did it affect the relationship between private and public space? It did so by making public space more public. Georg Simmel (1950) argued that one of the ways in which individuals dealt with the urban metropolis was to hide in the cloak of anonymity. In the crowds of strangers moving through the city, there was a privacy in numbers that was not present in rural areas where someone was likely to be recognized walking through a town. Drawing from Simmel, Turo-Kimmo Lehtonen and Pasi Mänpää (1997) argued the same with their idea of "street sociability," which suggests that in public in the city we are at once interested and yet indifferent and anonymous. We expect a certain amount of privacy even in public spaces, and we maintain that privacy by remaining anonymous faces in the crowd.

The portable camera challenges the anonymity that individuals use to maintain their privacy in public. As Warren and Brandeis (1890) argue, when someone's image is captured by strangers while she occupies public space, she is no longer anonymous. The level of privacy one expects even while in public is taken away, making the public even more public because one's image can later be shared with newspapers or interested others. The portable camera complicates the public/private dichotomy in ways that do not fit well inside Weintraub's four-part model. Instead, the portable camera shows that private and public are often subjectively experienced by the individual and can be more of a state of mind than a specific type of space. The crowded street can feel private for some people because they enjoy an anonymity not present in smaller towns or more closely knit neighborhoods. But image-capturing technologies can take away that anonymity, making the space feel more public and exposed for the person whose picture is taken and whose anonymity disappears. There are no lines that mark an experience as public or private; instead, a street can be both or neither depending on the perspective of the individual.

Interestingly, the portable camera works in contrast to how other mobile technologies such as the book, the Walkman, and the iPod are frequently analyzed as technologies that help individuals privatize public spaces. We, however, look at these technologies differently: We argue they help individuals control their experience of public space.

The Book as the Interface to Private and Public

In the last chapter we saw that before the nineteenth century, books tended to be large and immobile. Texts were supposed to be read aloud. As Jason Farman (2011) notes, "The process of reading is one of deep attention, focused between the text and the individual. Only when a text is read aloud does it enter the sphere of the exterior, group space. Otherwise, the text is ingested internally as a function between the eye and the individual" (p. 119). Reading became an individualized act and a private activity typically performed in demarcated sites, whether homes, monasteries, or libraries. By the nineteenth century, publishing houses began producing mobile, paperback novels in part because of the new social environments of new transportation infrastructure. These new environments of the railway can be understood in the terms of the sociability divide of private/ public. As we have seen, private has traditionally been associated with intimacy and controlled sociability. Before the train, most bourgeoisies traveled in carriages with a select few individuals with whom they were familiar. The railway changed transportation infrastructure from the relatively private space of the carriage to the more public space of the train compartment that was filled with strangers. To negotiate the new, more shared nature of travel, and feel like they had some control over the heterogeneity of the new social environment, people focused their attention on books and newspapers. The space of the train could not be made private in the same sense as the carriage, but by avoiding random sociability in the more public train compartment, individuals could feel more comfortable and familiar in those strange spaces (Figure 2.2).

Books and newspapers allowed readers to carve a more private experience out of the shared public space of a crowded street or a crowded compartment. In the last chapter, we looked at how people could use books and newspapers to control how they engage with public spaces. Reading in public allows individuals to manage the public nature of a place by controlling the social nature of public life. One could conceptualize this as privatizing the public space of the train, but that is too general a formulation. The person reading the book is still a part of the larger public space, even if that person is enjoying a momentarily private experience. Even if there is a subjective feeling of privacy in public through the immersion in a newspaper or novel, the subjective experience still exists in relation to the larger social situation. The reader still must position herself physically in relation to other people in the space and must recognize to some degree the dynamics of the shared space. This all points to the relational and negotiated nature of what is public and what is private. Susan Gal (2002) argues that "the distinction between public and private can be reproduced repeatedly by projecting it onto narrower contexts or broader ones" (p. 82). That is, some things can be considered

Figure 2.2 Passengers riding on trains often use newspapers and magazines to avoid random sociability. Picture taken in Berlin on May 29, 1958 on the subway line between Friedrichsfelde and Alexanderplatz. Copyright: Deutsches Bundesarchiv (German Federal Archive).

private in certain contexts, but public in others, which we saw with the example of the portable camera. Gal uses the example of a house: A house is typically considered a private space, but the rooms of the house are then defined as private or public in relation to other rooms; for example, the living room is generally understood as the public space of the house even as the house is private when compared to the park across the street.

The same can be said about a mobile technology such as a book. At the more macro level, a crowded street is a public space regardless of whether or not someone is reading or listening to a book. But people sitting on a bench on a crowded street reading a newspaper or a book may be seen as occupying a more private space in comparison to others in that space who are not using mobile technologies to tactically exert control over their experience. Once again, this type of experience shows the shortcoming of Weintraub's four-part model by emphasizing that the blurring of private and public often happens inside the mind of the individual. The book allows the individual to divert attention away from the public nature of the street and to create a somewhat controlled experience of the space through the text of the reading material. Note, however, that reading a book does not "destroy" public space by fully privatizing it. Instead, even if the public changes with different subjective experiences of space allowed by mobile technologies, the space is not less public. Rather, these subjective experiences, whether seemingly more private or not, are combined to form a different public space, consisting of different ways of self-presentation and co-existence. The subjective must still exist in relation to the inter-subjective nature of a place, and as we discussed in the beginning of this chapter, public spaces are spaces of co-existence and heterogeneity. A book or a newspaper does not negate that. Rather, the book becomes part of the experience of the public for the reader, not as a way to fully privatize or remove oneself from the public, but instead as a way to negotiate the (sometimes awkward) experience of being in public.

Books and newspapers, therefore, are also interfaces that individuals use to inform others about their levels of engagement. Ron Scollon (2001), drawing from Erving Goffman, argues that people who read in public demand a certain amount of civil inattention. By reading a book, they let others know that they are not fully or socially engaged with the public space. To address readers directly without violating norms requires an extra set of ritual interactions. For example, if someone is reading a book on a subway, the person sitting across from her would likely have to apologize before beginning a conversation. A productive way to think of these ritual interactions is to compare them to a room in a house. Before entering someone's room, it is customary to knock and ask permission. For the person who uses the book as an interface to a public space of a subway car, the apology required to strike up a conversation can be thought of as a knock on the door. The civil inattention required by reading in public means that to invade the mental space of the reader, the other must apologize and ask permission. Unlike with a physical wall in a room, however, readers may quickly re-engage with the broader public space if the social situation requires them to do so. As discussed in the last chapter, if the dominant involvement of the train passenger is brought to the foreground by, let's say, an announcement of the controller letting people know that the train will be delayed, the reader can quickly talk to the person next to him about how unfortunate it is that he will miss his appointment. So, rather than privatizing the space of the train, reading a book in public can be viewed as a form of "going away" (Goffman, 1963). That is, the book helps the reader to temporarily disengage from the space of the train by paying attention to something else. The book, in this case, works as a permeable boundary (interface) to that space, which changes how the reader experiences that space. It lets the reader control his experience of that space, enacting the type of control typically associated with private spaces. This does not mean that public and private spaces are a binary in direct opposition to each other. Rather, there are degrees of private in public and vice versa. As a result, the reader is still part of the larger, public situation of the train and can re-engage with others at any time by putting the book or newspaper down.

The Space of Sound and the Sony Walkman

If tapping a reader on a shoulder is similar to knocking on a door, then one might have to knock a little louder to get the attention of somebody with headphones in her ears. Headphones have been viewed as a technology that allows people to occupy a private sound world even when they are among groups of people (Bull, 2000; Hosokawa, 1987; Sterne, 2003). They are most often associated with mobile technologies, particularly the Sony Walkman and the Apple iPod. Long before the Walkman, however, people used headphones to interface the relationships they had with people in their surrounding space. What may be most interesting about early headphone use is that the technique necessary to occupy an individualized soundworld existed before it was materialized in the technology of the headphones.

To most of us, listening seems like something natural, something we are born knowing how to do. Much of what we consider natural now, however, is actually what Jonathan Sterne (2003) calls *audile technique*: We have to learn how to listen in specific ways. For example, in the eighteenth century, performances tended to be a shared experience where audiences interacted with the performers. Performances were mostly social events and the audience frequently intruded upon the performance by calling out to the performers and conversing with other members of the audience. It was not unheard of for the actors to have conversations with audience members while on stage. But in the nineteenth century, individuals learned how to listen to a performance as if it were a private experience. People no longer interacted with others; they instead carved out their own private soundspace and experienced the performance individually, ignoring the presence of others in the crowd. As Sterne describes, the development of audile technique made these performances more private. While people were still sitting in the public space of the concert or the opera house, being in a performance was viewed more as a personal (and therefore private) experience, rather than a public experience shared with other listeners. Listeners, therefore, created their own individualized spaces inside the larger shared space.

These audile techniques developed in Western society have transformed the acoustic spaces we inhabit. Sterne suggests that, by the nineteenth century, listeners were "alone together" in the private sound world created by audile techniques (p. 165). As we will see below, this idea of "alone together" is echoed in arguments about the privatization of space related to the Walkman and the iPod.

Sterne's description of learned acoustic spaces shows the difficulty of dividing public and private as separate entities. In one sense, the physical space of the opera house is certainly a public space; in another sense, however, Sterne describes people sharing a space while each engaged in a personal experience. This example shows how conceptions of private and public are subjective and often not clearly defined. The people in the opera house still occupy a public space filled with unknown others, but the nature of the public is altered by the individual experience of the acoustic soundworld. The combination of the two form a new public/private hybrid that might involve less conversation but is no less public or physically shared than a seventeenth-century performance. The lack of lively conversations among strangers does not mean that space has become private; rather, these individual experiences are part of the public space. The shared setting does not die; but these dynamics change the nature of public space as audile techniques become more prevalent and ingrained in everyday life.

Headphones are a sort of material instantiation of the individualized soundworld created by the audile technique honed in the nineteenth century. They are both a technique and a symbolic marker of personal space, as we discuss in more detail below. Headphones eliminate one element of the shared nature of the public—the auditory experience—because the sound played through the headphones is experienced only by the person who wears them. While headphones are now most commonly associated with mobile technologies such as the Walkman and the iPod, they were actually used for decades before the Walkman, mostly attached to more sedentary technologies. Telegraph operators wore headphones, and as the image in Figure 2.3 shows, they were able to concentrate on their work by blocking out the other sounds present in the space.

Headphones also accompanied the development of home radios, and listeners were able to plug them into the radio and listen to their music at home without disturbing others (Sterne, 2003). Notice that these headphones were attached to radios that were most often played inside the home. Just as the camera could make public space a little more public, headphones could make the private space of the home a little more private. A man could sit in a room with the rest of his family listening to music or news only he could hear. Often though, more than one person would sit together to listen to the same program as Figure 2.4 shows. Marketing materials for phonographs, for example, included images of people standing around together, listening to multiple pairs of headphones, staring down at the ground. Sterne (2003) argues that even though they were engaged in the shared experience of listening, they still remained distinctly separate from each other and their engagement was with the machine and not the others they were with. He claims, "Headphones isolate their users in a private world of sounds. They help create a private acoustic space by shutting out room noise and by keeping the radio sound out of the room" (p. 87).

However, the idea of a private world of sound needs to be complicated by thinking through what Sterne means by "private." As we have argued throughout this chapter, public and private are not static entities. Putting on headphones does not automatically disconnect somebody from their physical surroundings. We have seen that many urbanists and sociologists define good public spaces as spaces where strangers interact and socialize with each other (Gordon & de Souza e Silva, 2011; Whyte, 1980) However, as with the previous example of the opera house, the fact that people are not actively talking to each other while co-present does not mean that they are disconnected from that space, or that there is no interaction with it. For example, Figure 2.4 shows children listening to a Christmas radio concert through headphones. If we compare this to a situation where people are sitting around a table, listening to someone tell a story, chatting and laughing

Figure 2.3 A wireless operator typing a telegram at the Radiotelegraphic French coastal station of Saint-Mandrier-sur-Mer in 1995. Copyright: Centre d'instruction naval de Saint-Mandrier.

Figure 2.4 A radio concert for children in the streets of London during Christmas Eve, December 1930. Note the collective use of headphones.
Copyright: Deutsches Bundesarchiv (German Federal Archive).

together, we might think that the children are separate, isolated from each other. But if we look at the situation differently, we might suggest that they are sharing the experience of listening to the radio, and at any time they can remove the headphones and share what they are hearing. So the "private world of sound" is not a complete privatization of the shared space, nor does it necessarily isolate users from each other.

When the first mobile music players were introduced, headphones became even stronger interfaces to public and private spaces. As we have seen in the previous chapter, the first personal mobile listening device was the Sony Walkman. With the Walkman, what Sterne calls "private acoustic space" could be transported with the user. The Walkman user still occupies a shared, public space, but she does so differently. She replaces the sounds of the city with an individualized soundscape that just belongs to her (Hosokawa, 1984). The Walkman then becomes an interface that shapes the individual experience of public/private space.

The Walkman was a phenomenon. As du Gay et al. (1997) argue, it was one of the first technologies that influenced the popular opinion of post-industrial globalization because it introduced the world to the image of "Japanese-ness." Built in 1978 for Sony co-chairman Akio Morita, the Walkman was then marketed in both the United States and Europe. The marketing of the Walkman is what du Gay et al. focus on most extensively, arguing that by crafting advertisements that were at once targeted at a mass audience and individualized, Sony achieved "the best of all possible worlds—mass marketing and personal differentiation" (p. 31). The personal differentiation came from how Sony sold the Walkman as a "hip" new device, both targeting and constructing audiences at the same time. Walkman advertisements featured beautiful young people enjoying the freedom of mobile listening practices.[2] Sony sold the Walkman as a technology of freedom for the young. They could block out the outside world, but at the same time show the outside world their status as a cool young person.

A notable design feature of the earliest Walkman model was that it included two headphone jacks. The two jacks allowed the owner to listen to music with a friend, similar to the images of the early radio discussed above. The Walkman designers thought the addition of the extra jack was important because they did not think many people would want to listen to music alone in public spaces. They thought people would see this as rude and undesirable. But the early designers were wrong: People did prefer to use the Walkman as an individual, rather than a shared technology. du Gay et al. (1997) point to the removal of the second headphone jack as an example of consumers' influence over the design process of a technology, a change that carried through to the design of CD players and now iPods.

The unpopularity of the second headphone jack says a great deal about how people understood the purpose of the Walkman. The Walkman was a personal form of media, one that individuals wanted to experience by themselves. While some might highlight the ability to share the personal stereo by using one earbud while a friend uses another (Hosokawa, 1984), that is relatively rare. The Walkman works as a personal interface to public spaces that individuals use to control their experience of the public. Like with the book, this control does not lead to a complete privatization of the public, but instead it foregrounds the complex relationships between the two because people use the technology as a means of managing the co-present situation by listening to music. Walkman users feel more in control of the public space around them, a feeling generally associated with private spaces.

Nonetheless, some critics of the Walkman have argued that it led to the privatization of public space (du Gay et al., 1997). As Sterne discusses in the context of earlier auditory technologies, headphones represent a type of interior listening that is private, which has led to major questions about the Walkman's effect on public space. Critics asked whether a public space was still public if everyone was listening to their own music (Bloom, 1988; du Gay et al., 1997). They feared that public space would be turned into the equivalent of a huge reading room, with everyone sitting in a row, heads down, reading different material, and losing the awareness of the shared experience. For example, Alan Bloom, the great protector of the "traditional" canon of great Western literary works, criticized the Walkman for contributing to an atomized society of individual actors who did not experience the shared traditions of society. Bloom (1988) wrote, "As long as they have their Walkman on, they cannot hear what the great tradition has to say" (p. 81). Bloom was not alone in criticizing the Walkman for its effects on society. Describing these criticisms, du Gay et al. (1997) write that the Walkman

> Triggers off many themes associated with late modernity as a distinctive way of life: The lonely figure in the crowd, using the media to screen out the routines of boring, everyday life; the emphasis on mobility and choice; the self-sufficient individuals wandering alone through the city landscape— the classic Walkman person seen so often in its advertisements, the urban nomad.
>
> (p. 16)

The larger question to which du Gay et al. point is whether the public life of the streets can even be considered public if it is filled with mobile individuals filtering out public sounds with an individualized soundscape. As we have seen in the last chapter, the Walkman does represent a greater trend to turn away from public life and toward the private experience that Sennett identifies. But these criticisms reveal a certain technological determinism: When they criticize the Walkman for its negative effect on public life, they fail to identify the trends Sennett had already recognized, that is, the decline of urban sociability, and the private encroaching upon the public. The "impersonal codes of conduct" that had long defined public life had already begun to wither; the use of Walkman in public was a reflection of that trend.

Whether public space is adulterated by the experience of mobile technology use, however, depends on the definitions of public and private one is using. In the citizenship model closely associated with Arendt (1958) and Habermas (1989), the public is the site of political deliberation and democracy. If we adopt that version of the public, then yes, the public does suffer if people are using mobile technologies. Someone with headphones in her ears is unlikely to debate important issues with the other people she is sharing that space with or be exposed to differing viewpoints. As we discussed in the previous chapter, people often use mobile technologies to interface the way in which they engage with others, and that often results in using them to avoid socializing with others. From the citizenship perspective, the lack of engagement with others may represent a full-scale invasion of the public, political realm by the private. However, as we pointed out earlier, the citizenship model is only one way to understand the public/private continuum. According to the sociability model, for instance, public spaces are spaces where strangers congregate. Although strangers share the same public space, they do not necessarily socialize with each other.

Earlier in this chapter we discussed how so many of the ways in which people traditionally divide the private from the public have been blurred into hybrids. For example, Western societies have an economy that has blurred the private of the market with the public of government. The public sphere in a political sense is often now enacted from the domestic, private space of the home through Internet chat rooms. When we understand the public/private continuum not in absolute fixed terms but instead attend to the complex intricacies of their always shifting boundaries, it is hard to support a wholesale takeover of the public by the private; consequently, Walkman use cannot fully privatize public space. If public spaces are spaces filled with strangers and difference, then using mobile technologies in public does not make that space private because the spaces are already filled with strangers with whom people do not intend to socialize. They do, however, allow individuals to exert types of control over experience typically associated with private spaces. It is this combination of control and co-present heterogeneity that leads to private-in-public hybrids.

More than any other scholar, Michael Bull (2000, 2001, 2004b) has examined the different ways in which individuals use the Walkman to control their experiences of public spaces. He argues that the Walkman allows for a more advanced form of what Raymond Williams (1975) calls "mobile privatization." As we discussed above, Williams initially used the term to refer to broadcast media and one of the great paradoxes of Modern society: "An at once mobile and home-centred way of living" (p. 19). Through the broadcast medium of television, individuals could experience increased virtual mobility even as life became more and more centered in the private space of the home. Bull (2004a) disagrees with Williams' definition of mobile privatization, arguing that Williams relies far too much on a visual epistemology. For Williams, whether the mobile privatization occurs through the screen of a television or the windshield of a car (Williams, 1988), it is typically the act of looking that is held up as primary. Bull argues that using sound technologies to move through the city is a much more literal form of mobile privatization. Williams discusses how people withdraw into "shells," whether they be tight-knit family units or comfortable automobiles. According to Bull, for the Walkman user that "shell" is constructed on a more personal level by the soundtrack she chooses to compose her experience of the spaces she moves through. Bull's analysis, however, too strongly emphasizes the death and disintegration of public spaces caused by the use of mobile technologies. He writes that for many Walkman users "Public spaces are voided of meaning and are represented as 'dead spaces' to be traversed as easily and as pleasurably as possible" (p. 79). He later says that "Public space in this instance is not merely transformed into a private sphere but rather negated so as to prioritize the private" (p. 79).

Concepts such as mobile privatization suggest that mobile technology use in public space empowers users to transform public space into a private setting, and that this leads to the death of public spaces. We do not deny that using a Walkman helps people feel more familiar with public spaces by experiencing a certain form of control over that space. But they are still using the technology in public, in a place full of strangers, unpredictability, and anonymity (Lehtonen & Mänpää, 1997). In order to deal with the heterogeneity of public spaces, users seek to colonize them through the soundtrack of their music in part as a way to interface with the public on their own terms. The desire to mediate our relationship to public spaces existed before the Walkman, and that desire will not disappear anytime soon. Acknowledging this desire for control and the always mediated experience of public spaces is very different from the privatization of space that Bull uses as a lens through which to analyze Walkman use.

Bull is only operating from one conceptualization of the public, the conceptualization that "good" public spaces are spaces where there is active co-present interaction. As we have seen in the last chapter, this view is not new. Urban scholars such as William Whyte (1980) have already highlighted the importance of interaction among strangers as one of the desirable components of good urban spaces. But in reality, people are often anonymous and fairly indifferent in public spaces. If we think of the public as the site of heterogeneity and co-existence rather than just a site of actual co-present interaction, then the public is not negated by Walkman use. Rather, Walkman use becomes part of the fabric of public spaces, and indeed an interface that helps users manage their interactions with the public. Walkman users still belong to public spaces, even though they might seem indifferent and anonymous. We saw earlier in this chapter in our discussion of Lehtonen and Mänpää's (1997) concept of street sociability, in which they argue that anonymity and indifference are necessary to survive the stimulation of urban spaces. Few would argue, however, that people who remain anonymous do not occupy a public space. Rather, following Georg Simmel (1950), anonymity is a required component of urban spaces. The individual anonymity, along with all types of psychological and technological *blasé* attitudes, are necessary interfaces to participate in the public. They construct it, rather than eliminate it. They make the experience of public space desirable in a way that is substantially different from just transforming it into a private space. The public then can be understood as a space that may include many individuals who are engaged in more private activities and not actively interacting with each other, but it is not "voided of meaning." The meaning of public instead is shifted as it is made more complex by the "private-in-public" activities of individuals. The Walkman and the iPod, as we discuss below, work much the same way. People listening to headphones still occupy public space, even if it is not the idealized type of public space that some critics envisioned.

From Sony to Apple

Almost everything we have said about the Walkman can also be applied to the Apple iPod and other MP3 listening devices. Like the Walkman in the 1980s, the iPod has become one of the defining cultural icons of our time (Coleman & Gøtze, 2001), even though it has begun to be replaced in the cultural consciousness by smartphones. The iPod certainly has more processing power than the Walkman, but it provides a similar experience of public space. Individuals still use them as a tool for civil inattention, just like they did with the book and the Walkman. Like the Walkman, iPods also allow users to take a practice previously confined to specific places (listening to music) into the streets of the city. The shift with the iPod is the larger database of music that users are able to carry with them and the increased functionality, which gives them the ability to more precisely match their favorite soundtrack to the space they move through. With the Walkman, one's portable music library was restricted to the number of cassette tapes one carried, but the iPod lets the listener access an entire music library while mobile. The ability to control the choice of music is not a minor shift in understanding how iPod use complicates the boundaries between private and public.

Discussing Walkman use, Bull (2000) introduces the idea of filmic experience, in which one engages with public space as if other people are acting out roles to match the soundtrack playing through one's Walkman. Extending that idea, it is possible to think of the iPod as a remote control to the filmic experience. With filmic experience, the listener views the public life as if people are acting out roles in a movie, with the music playing through the headphones as the soundtrack to the actions of the other. No one can change the channel like on television and introduce an entirely different cast of characters, but by changing the music to the film, iPod users can change their experience of the public space, personalizing it even further. Note that through the filmic experience, iPod users are not disconnecting from the public space around them, but actually experiencing that space in a different, more controlled way. The people and things present in the physical space are necessary elements of the filmic experience, just as the music is. The elements of the physical space cannot be changed, but the music can. So for someone walking by a library, a couple engaging in a private conversation under a tree can seem very different depending on the song the listener chooses to play on her iPod. A love song can make the couple seem romantic, a break-up song can make them seem tragic. The couple remains under the tree regardless of what the iPod user does, but the iPod user can choose how she views the couple through the music she feels is most fitting to the experience.

Because of the larger size of its music database, the iPod was marketed as empowering users to control their experience of public spaces in ways not possible before. This theme of control is represented in the influential advertising campaign used to promote the iPod. More than maybe any other tech company, Apple has created advertisements that are immediately recognizable and memorable (Stein, 2002). While Bull (2000) can argue against the importance of Sony's early Walkman ads, there is little doubt about the importance of Apple's iPod campaign, which received the Media Lion Award for best international advertising campaign and the EFFIE award, the highest award bestowed by the advertising community (Cooper, 2009; Jenkins, 2008).

The early iPod advertisements clearly focused on the idea that mobile technology use detaches people from the space they move through by depicting iPod users as a "silhouette" over an empty (although colorful) background. This now iconic "silhouette" campaign became so popular that it was later parodied extensively by both competing companies and anti-war groups (Jenkins, 2008). Troy Cooper argues that the iPod advertisements extend the free-thinking ethos Apple constructed in the Macintosh 1984 advertisement by promoting individuality and freedom and are effective in part because of their simplicity. They also represent the idea that mobile technologies can lead to a fully private experience of space. They feature five basic elements: "the dancing silhouettes, a uniform neon backdrop, upbeat music, the white iPod, and a small amount of text and logo" (Cooper, 2009, p. 475). These five elements combine to emphasize the freedom of the dancing figure who uses the earbuds to carve out a private space, moving to the beat through the headphones as if nothing else matters. The iPod ads clearly define it as an individualizing technology, one that helps users to bear and control the otherwise chaotic experience of public spaces (Figure 2.5). Eric Jenkins' (2008) description, in his article about the iconography of the silhouette campaign, is worth quoting at length:

> The experience of watching the ads, when engulfed in the 30-second moment, simulates the experience of the world through headphones. Anyone who has traversed public space while entranced in their favorite

Figure 2.5 A 2006 Apple iPod billboard in San Francisco, California. Copyright: Creative Commons license. Picture taken by Naotake Murayama.

song recognizes the experience, similar to the feeling one gets when consumed in dance. The world seems to become mute, while people appear to move in harmony with your song. The listener experiences an immediate and total noise that emanates from the inside, shutting out the panauditory experience of proliferating noise common in contemporary life. The brilliance of Apple's iconic portrayal is in bringing together so many associated elements of this common phenomenological experience. The dancing, the rhythmic music, the headphones, and the neon backdrop all reference the experience of immersion in music.

(p. 477)

For our analysis here, the immersion in music is what might be the most important part of the Apple advertisements. The silhouette advertisements showed dancing figures that literally exist in their own worlds of sounds. Through the music, they are able to detach themselves from the places they move through by carving their privatized public space, thus making it their own. The backgrounds emphasize the privacy of the experience; they are stark in their neon emptiness, highlighting the dark solitary figures alone in a world they are constructing through their movement and the music. Apple's silhouette campaign certainly positioned the iPod as a "cool" product, similar to the Walkman campaign, but the freedom of the silhouettes take the emphasis a step further, showing a world in which the listener is alone—the context of the city, an integral part of many Walkman ads, is absent from the iPod commercials. The act of dancing like no one is watching is the epitome of a private experience of space—one that, as we have shown, is common in discussing mobile technology use. By deciding to market the iPod as a technology that detaches users from their surrounding space, Apple tapped into existing anxieties about co-existing with strangers in urban centers and offered users a way of controlling their interactions with strangers and the spaces they move through. Just as Bull (2000) argued that for some Walkman users "public spaces are voided of meaning" (p. 79), Apple's advertisements literally void public space by making it a solid, empty background.

The idea of detachment from space represented by the silhouetted dancers of the iPod commercials was also echoed by some members of the popular press. Warren St. John (2004), writing for the *New York Times*, wrote an editorial about the invasion of the "iPod people," comparing them to zombies in a horror movie. He writes about iPod owners walking through the city in a daze, standing in lines with no awareness of their surroundings. They are more docile forms of the iPod silhouettes, not dancing around, but interiorized through the music blaring through their headphones. St. John's criticism is extreme, and as we have been discussing in this book, individuals used different techniques to manage their experience of the stimulation of the city long before the iPod. The iPod is just a new way to do so, letting individuals interface with public space by laying an individualized auditory layer over the space. There is still a certain amount of co-presence and shared experience when inhabiting a public space, contrary to the empty backgrounds of the silhouette ads. To function in that shared space, the individual has to have at least some recognition of the larger social situation, regardless of what is blaring through the headphones.

The Meaning of Public Spaces

As with the book and the Walkman, the iPod has two main roles: It allows individuals to take activities that were previously confined to what is often considered private space, such as listening to music, out into the shared space of the public, and it helps individuals construct a more personalized experience of public space. These two aspects of iPod use have led to the questions about the death of public space. These questions are based on the implicit assumption that for a public space to be healthy and vibrant, individuals must be interacting in a co-present manner. But have large cities ever really had streets where strangers freely and willingly interact? Simmel argued that the increasing size of urban spaces led to a lack of interaction with strangers. Mobile technologies, then, are not simply a cause of this lack of interaction, but rather are also a consequence. Like the *blasé* attitude, they are new types of interfaces to public spaces. Reading a book or listening to an iPod are activities that generally do not directly involve others in a shared space, but perspectives that claim that these activities simply privatize public spaces miss the nuances and possibilities of the ongoing existence of private-in-public, and the constantly shifting nature of the boundaries between the private and the public. More importantly, they miss the perspective that people often expect to have a certain degree of privacy in public spaces, as well as anonymity.

To exist in public space, whether we are talking with strangers or listening to music, requires a great deal of what Robert Putnam (2000) calls *thin trust*, which is a generalized trust in others to do what they are supposed to do. Public space cannot exist without a great deal of thin trust. As Gordon and de Souza e Silva (2011) write, "public space is a collection of minor social contracts" (p. 90). We expect other people to hold up their part of the contracts, and we will hold up ours. For example, walking down the street requires a great deal of trust in others. We trust that cars driving past will not run up onto the sidewalk; we trust that other pedestrians will not shove us into the street; we trust that the man on his mobile phone is not using it to activate a bomb. Public space is made up of these different layers of thin trust whether we are directly engaging in a co-present manner with others or not. When people use mobile technologies to avoid engaging with others, public space still does not become privatized because occupying public space requires implicitly signing on to all these minor contracts. Most people reading books or listening to music remain aware enough of the public nature of the space to live up to the general level of thin trust that is required of them.

In addition, we must acknowledge that we have seen even more significant shifts in the nature of the public with the advent of online communities. The public must be more than people engaging in co-present interaction or else it would be difficult to conceptualize how online communities formed around personal social networks could be considered publics. Many of these "networked publics," as danah boyd (2007) calls them, do not feature the levels of heterogeneity desired by some of the thinkers we have outlined in these two chapters and they are a more controlled experience than walking down a busy city street filled with strangers; however, these networked publics are new forms of publics in which people learn how to engage with others and practice social skills. They also involve the layers of thin trust that are necessary for the existence of physically located public spaces. In her analysis of the networked publics of social networking sites, boyd discusses the changing dynamics of the publics many of today's teens grow up in. Teenagers are often restricted from enjoying public spaces in the ways they desire, and they instead participate in social life through wall posts, private messages, and status updates. As boyd (2007) writes, "by doing this, teens are taking social interactions between friends into the public sphere for others to witness" (p. 124). Social networking sites offer these teens their best opportunity to exist in an unregulated public, free from the gaze of adults, and it is through these public experiences that many people live their social lives. This is a more controlled form of engaging with the public, just as an iPod user engages in a more controlled experience of shared space, but it still must be considered the public for the concept to retain much meaning. People still must rely on trust and there must be expectations of how others will behave, just like the publics of the city. Like with mobile technology use, networked publics represent the changing nature of the public, not the death of it.

Now, with the emergence of dialogic mobile technologies, such as mobile phones and location-aware technologies, we witness yet another change in the way we understand public spaces. It is as if the idea of networked publics is brought into physical spaces, redefining the borders between online and face-to-face communities, digital and physical spaces. As we will see in the following chapters, physical spaces become clustered with networked connections, in what Gordon and de Souza e Silva (2011) call *net locality*. Contemporary accounts of urban public spaces need to take into consideration both face-to-face and remote connections. In this new logic of public spaces, "the purview of what is near has expanded beyond that which is right next to you, and paying attention to an anonymous user at a neighboring street corner, visualized on a mobile map, is just as likely as paying attention to the stranger across the street" (Gordon & de Souza e Silva, 2011, p. 86). The idea that both remote and co-present interactions are now interfaced via mobile technologies fundamentally redefines how we understand public spaces and the character of locations.

Notes

1 Private spaces are only controlled spaces in comparison to public spaces. Inside private spaces, people have differing levels of control, as feminist theory has so productively shown.
2 For a sample of Sony's early Walkman advertisements, see the following sites: http://www.grayflannelsuit.net/retrotisements-rip-sony-walkman-1979-2010/ and http://gizmodo.com/5305177/great-sony-walkman-tv-and-print-ads-of-the-1980s.

References

Arendt, H. (1958). *The human condition*. Chicago: University of Chicago Press.
Aristotle (2010). *Politics* (W. Ellis, Trans.). London: Lits.
Bimber, B., Flanagin, A. J., & Stohl, C. (2005). Reconceptualizing collective action in the contemporary media environment. *Communication Theory, 14*(4), 365.
Bloom, H. (1988). *The closing of the American mind*. New York: Simon and Schuster.
Bobbio, N. (1989). *Democracy and dictatorship*. Minneapolis: University of Minnesota Press.
boyd, d. (2007). Why youth love social network sites: The role of networked publics in teenage social life. In D. Buckingham (Ed.), *Youth, identity, and digital media* (pp. 119–142). Cambridge, MA: The MIT Press.
boyd, d. (2008). Facebook's privacy trainwreck. *Convergence: The International Journal of Research into New Media Technologies, 14*(1), 13–20.
Bull, M. (2000). *Sounding out the city: Personal stereos and the management of everyday life*. Oxford: Berg.
Bull, M. (2001). The world according to sound: Investigating the world of Walkman users. *New Media and Society, 3*, 179–197.
Bull, M. (2004a). Sound connections: An aural epistemology of proximity and distance in urban culture. *Environment and Planning D, 22*, 103–116.
Bull, M. (2004b). Thinking about sound, proximity, and distance in Western experience: The case of Odysseus's Walkman. In V. Erlmann (Ed.), *Hearing cultures: Essays on sound, listening, and modernity*. New York: Berg.
Bull, M. (2007). *Sound moves: iPod culture and urban experience*. New York: Routledge.
Burrows, R., & Ellison, N. (2004). Sorting places out? Towards a social politics of neighborhood informatization. *Information, Communication and Society, 7*(3), 321–336.
Castells, M., Linchuan Qui, J., Fernández-Ardèvol, M., & Sey, A. (2007). *Mobile communication and society: A global perspective*. Cambridge, MA: MIT Press.
Coleman, S., & Gøtze, J. (2001). *Bowling together: Online public engagement in policy deliberation*. London: Hansard Society.
du Gay, P., Hall, S., Janes, L., Mackay, H., & Negus, K. (1997). *Doing cultural studies: The story of the Sony Walkman*. London: Sage.
Farman, J. (2011). *The mobile interface of everyday life: Technology, embodiment, and culture*. New York: Routledge.
Foucault, M. (1999). Space, power and knowledge. In S. During (Ed.), *The cultural studies reader* (pp. 335–347). London: Routledge.
Gal, S. (2002). A semiotics of the public/private distinction. *Differences: A Journal of Feminist Cultural Studies, 1*(1), 77–95.

Giddens, A. (1991). *Modernity and self-identity: Self and society in the late modern age.* Stanford, CA: Stanford University Press.

Goffman, E. (1963). *Behavior in public places: Notes on the social organization of gatherings.* New York: Free Press of Glencoe.

Gordon, E., & de Souza e Silva, A. (2011). *Net locality: Why location matters in a networked world.* Boston: Blackwell Publishers.

Habermas, J. (1989). *The structural transformation of the public sphere.* Cambridge, MA: MIT Press.

Hosokawa, S. (1984). The Walkman effect. *Popular Music 4,* 165–180.

Hosokawa, S. (1987). *Der Walkman-Effekt.* Berlin: Merve Verlag.

Huhtamo, E. (2004). Hidden histories of mobile media. *Vodafone Receiver, 11,* 1–4.

Jacobs, J. (1961). *The death of life of great American cities.* New York: Random House.

Jenkins, E. (2008). My IPod, my icon: How and why do images become icons? *Critical Studies in Media Communication, 25*(5), 466–489.

Lehtonen, T.-K., & Mänpää, P. (1997). Shopping in the East Centre Mall. In P. Falk, & C. Campbell (Eds.), *The shopping experience* (p. 136). London: Sage Publications.

Marvin, C. (1988). *When old technologies were new: Thinking about electric communication in the late nineteenth century.* Oxford: Oxford University Press.

Peters, J. D. (1999). *Speaking into the air: A history of the idea of communication.* Chicago: University of Chicago Press.

Putnam, R. D. (2000). *Bowling alone: The collapse and revival of American community.* New York: Simon & Schuster.

Rosaldo, M. Z. (1974). *Woman, culture, and society.* Stanford, CA: Stanford University Press.

Scollon, R. (2001). Action and text: Toward an integrated understanding of the place of text in social (inter)action. In R. Wodak, & M. Meyer (Eds.), *Methods of critical discourse analysis* (pp. 139–182). London: Sage.

Sennett, R. (1977). *The fall of public man.* New York: Knopf.

Shapiro, S. (1998). Places and spaces: The historical interaction of technology, home, and privacy. *The Information Society, 14*(4), 275–284.

Sheller, M., & Urry, J. (2003). Mobile transformations of 'public' and 'private' life. *Theory, Culture, and Society, 20*(3), 107–125.

Simmel, G. (1950). *The sociology of Georg Simmel* (K. Wolff, Trans.). New York: Free Press.

Solove, D. (2008). *Understanding privacy.* Cambridge, MA: Harvard University Press.

Sterne, J. (2003). *The audible past: Cultural origins of sound reproduction*: Durham, NC: Duke University Press.

St. John, W. (2004). The world at ears' length. *New York Times.* Retrieved from http://query.nytimes.com/gst/fullpage.html?res=9C05E7DA1E3AF936A25751C0A9629C8B63&scp=1&sq=world%20at%20ears'%20length&st=cse.

Warren, S., & Brandeis, L. (1890). The right to privacy. *Harvard Law Review, 4*(5).

Weintraub, J. (1997). The theory and politics of the public/private distinction. In J. Weintraub, & K. Kumar (Eds.), *Public and private in thought and practice: Perspectives on a grand dichotomy* (pp. 1–42). New York: University of Chicago Press.

Whyte, W. (1980). *The social life of small urban spaces.* Washington, DC: Conservation Foundation.

Williams, R. (1975). *Television: Technology and cultural form.* New York: Schocken Books.

Wolfe, A. (1997). Public and private in theoretical practice: Some implications of an uncertain boundary. In J. Weintraub, & K. Kumar (Eds.), *Public and private in thought and practice: Perspectives on a grand dichotomy* (pp. 182–203). New York: University of Chicago Press.

3

FROM VOICE TO LOCATION

Johanna wakes up the following day feeling rested and relaxed. The sun is shining outside the window and she feels like going for a walk to get to know her new neighborhood. She walks about three blocks and finds a beautiful park, full of flowers and shadowed alleys. Because it is Sunday, the park is full of people: families with children, couples, and others by themselves. At the end of the main alley, she finds a small area with benches and tables. Most benches are occupied by groups of people, except one, on which a girl sits alone. Imagining that the girl won't mind, Johanna sits on the far end of the bench. She sits in the sun, closes her eyes, and just as she is almost relaxing, she hears the girl begin to talk to somebody on her mobile phone. The girl is apparently scheduling a meeting, or maybe asking the person at the other end of the line why she or he is late. Almost involuntarily, Johanna starts paying attention to the conversation. She is actually getting interested in the conversation, trying to imagine what has happened and who is at the other end of the line. At the same time, she looks around and sees that the man sitting at the bench right behind them has an annoyed look on his face. He is reading a book. "People always get annoyed with others talking on the phone in public spaces," she thinks. But before Johanna is able to refocus on the girl's conversation, the girl's phone starts to ring! The girl, caught by surprise, doesn't know what to do for a moment, but eventually answers the phone. While (now really) talking to the other remote person, she looks around, somewhat embarrassed, and walks away. Johanna starts laughing to herself: It is not the first time she has seen people fake a mobile phone conversation in a public space in order to show others they aren't "alone."

> The man reading a book is probably happy now. She looks at him again and sees a cup of coffee in his hands, which immediately sparks her craving for a good cup of dark coffee. "I wonder if there's a coffee shop nearby." She reaches for her phone and opens an app on her iPhone called *AroundMe*. She selects "coffee" and immediately five coffee shops pop up on the list. She walks to the closest one.

The mobile phone has probably received more attention than any other contemporary technology when it comes to discussing interactions in and with public spaces. Claims of the disruption of public spaces by personal conversations have been frequent in media discourses and in scholarly works that analyzed the social consequences of mobile phone use (Katz & Aakhus, 2002; Ling, 2004, 2008; Plant, 2001). So far we have been discussing the nature of public spaces and how the use of mobile technologies both changes them and also reflects already existing tendencies and patterns of sociability in these spaces. We purposely left the mobile phone out of this preliminary discussion because it adds two very important elements to this discussion that complicates the definition of public spaces: remote connection and location-awareness. As mobile technologies, mobile phones share many characteristics with other personal mobile technologies. Like the book, the Walkman, and the iPod, mobile phones have been claimed to privatize public spaces when individuals engage in personal conversations with distant others. The mobile phone user's focus on absent others was often associated with fears of disconnection from places. The argument went like this: If people were talking on the phone, they were probably more engaged with a remote person than with the co-present situation. In addition, because mobile phones are not attached to specific places (like landlines), mobile conversations were also generally associated with freedom from place.

The relationship between mobile phones and places has always been controversial. Many mobile phone scholars focused on how mobile phones transformed our perception of place, but they did so in ways that emphasized the decreasing importance of places and locations—and often also the death of public spaces. These studies focused primarily on how mobile phones helped users forge connections with remote others at the cost of forging social relationships with co-present people (Gergen, 2002, 2008, 2010; Hampton, Livio, & Sessions, 2010; Moores, 2004; Plant, 2001; Puro, 2002), and they focused on the use of mobile phones as two-way communication devices. Early locative media artists and locative media theorists are among the first to study how mobile technologies in general and mobile phones in particular change our experience of place and may be used to interact with people and information nearby (Galloway, 2006; Hemment, 2006; Hight, 2006; Paulos & Goodman, 2003; Tuters & Varnelis,

2006). With the increasing popularization of location-aware technologies, the relationship between mobile technologies and places requires a new perspective because these technologies strongly influence how people interact with their surrounding space and how they understand location. The automatic detection of location becomes a relevant interface that mediates users' relationships to places and among each other.

The goal of this chapter is twofold. First, we discuss how the relationship between mobile phones and place has been framed, focusing on a line of thinking that argues mobile phone use disconnects people from places. Second, we outline a brief history of location-aware technologies through locative media art and location-based services (LBS) to show how location-awareness brings place (and more importantly, location) to the forefront of users' interaction with information and other users. We begin by situating the literature on mobile phone use in public spaces within our earlier discussion of individualization and control over public spaces to show how these technologies also operate as sorts of selective interfaces to the public. We acknowledge a wealth of work that deals with how mobile phones challenge the borders between public and private spaces (Fortunati, 2002; Gergen, 2002; Geser, 2004; Ling, 2008; Puro, 2002), and we also show how this invasion of the public by the private has often been perceived as disconnecting users from their surrounding spaces. But contrary to most of these studies, we claim that individuals do not necessarily use mobile phones to disconnect themselves from their physical surroundings. As shown in the previous two chapters with the examples of the book, the iPod, and the Walkman, people use mobile technologies to actively choose when and where to engage with their co-present situation. The mobile phone complicates this issue because they are telephones, that is, they are used to interact with someone who is remote, which emphasizes the perception of disconnection from the co-present.

Until the end of the last decade, mobile phones were used almost exclusively for remote two-way communication. Mobile phones have always been location-aware, that is, they could be located by triangulation of waves, but the use of mobile phones as location-aware technologies has been restricted to the domain of art and research, and mostly ignored by mobile communication scholars. This scenario changed with the commercial popularization of third generation (3G) GPS-enabled phones. In the second half of the chapter, we address the shift that emerged with the popularization of location-aware phones, which emphasizes the relevance of location as an interface for social interactions. This chapter thus also explores how the emergence of location-aware mobile technologies not only redefines people's connections to places, but also redefines the character of locations, which acquire new meanings because they now become embedded with location-based information.

We outline a brief history of locative mobile applications, focusing on two time periods: First, we look at locative media artists, researchers, and early start-up companies that pioneered the development of location-based applications right after the removal of the Global Positioning System (GPS) signal degradation in 2000. Second, we describe the rapid growth of LBS—commercial services that use the phone's location to provide users with place specific information—after the commercial release of location-aware 3G phones (a.k.a. smartphones), marked by the iPhone 3G, Nokia's Symbian phones, and phones featuring Google's Android operating system. LBS include a diverse set of applications that help individuals navigate, find restaurants and other services, and read about the history of a place. As a subset of LBS, location-based social networks (LBSNs) and location-based mobile games (LBMGs) are social networking software that allows users to see the location of other users. In summary, we contrast early scholarly discussions of how mobile phones disconnected people from places with the parallel development of location-aware applications that explicitly mediate the relationship between users and locations.

From Mobility to Communication

The mobile technologies described earlier in this book allowed individuals to manage their relationship to place by helping them alternate their attention between a stronger engagement with the public and a stronger engagement with personal activities in public spaces. For example, somebody reading a book in a train may choose to chat about the book's topic with the person sitting beside him or may choose to ignore other events going on in the train compartment and focus on the reading. We thus have showed how these mobile devices help users manage their interactions with others in their surroundings by signaling different types of availability for face-to-face communication. The book, the Walkman, and the iPod, however, are not dialogic communication devices. Unlike mobile phones, they do not provide users with the ability to engage in conversations with distant people while moving through public spaces.

As noted by James Carey (1983), the emergence of the telegraph disconnected communication and transportation for the first time. Until the invention of the telegraph, communication was invariably connected to transportation, as the slow mode of horseback travel or the faster railway were required for the delivery of long-distance communication—that is, written letters. With the telegraph, however, messages could be sent to distant others without the need for physical movement. Nevertheless, while moving physically, whether on a horse, or on a train, physical travel has meant a disconnection from remote places. Travelers could communicate with others in the train compartment, but they could not communicate with somebody else across the country while on the move. Mobile phones and wireless radios changed this relationship between mobility and communication by enabling people to communicate with others while moving physically (Farley, 2005; Ling, 2004; Ling & Yttri, 2002). They combined the nearly instantaneous connection of the telegraph with physical mobility. But rather than being used in the interior of a car or a train, like early wireless radios and car phones, mobile phones could be carried around in public spaces.

Because of this public display of remote connection in a space where people are expected to interact with each other and pay attention to their surroundings, mobile phone users are frequently accused of isolating themselves (Campbell, 2008; Hampton et al., 2010). This is why early mobile phone studies differ from similar analyses of the fixed-line telephone. Several scholars who studied the emergence of the fixed-line telephone focused on its ability to help people overcome isolation from the outside world and increase local contacts with the surrounding community, particularly in rural areas (Fisher, 1994; Umble, 2003; Wiley & Rice, 1933). For example, Kenneth Gergen (2002) notes that the emergence of the fixed-line telephone increased the potential for face-to-face interactions, and John Durham Peters suggests (2008), "rather than eviscerating local life, cars and telephones actually multiplied the intensity of contacts" (p. 34).

However, Gergen (2002) does suggest that the telephone demands "that participants divorce their attention from their immediate surroundings" (p. 237). This was not often viewed as a problem though, because landline telephones were mostly used inside the private space of the home and the office. The problem arises when this remote communication is brought to public spaces. Studies of the mobile phone tended to examine how connections with remote others in public spaces detached individuals from their local environment. For example, when analyzing the use of mobile phones in Finland, Jukka-Pekka Puro (2002) suggested that "talking on the mobile phone in the presence of others lends itself to a certain social absence where there is little room for other social contacts. The speaker may be physically present, but his or her mental orientation is towards someone who is unseen" (p. 23). Puro's claims resembles William Whyte's (1980) and Jane Jacobs' (1961) idea of good and vibrant public spaces. As we have seen in the former chapters, vibrant public spaces have been conceptualized as shared spaces full of people actively chatting and interacting with each other. People by themselves in plazas or cafes are obviously part of the public place, but a place achieves its full potential as public when strangers interact with each other (Whyte, 1980). Puro suggests that because of the mobile phone people now prioritize remote connections at the expense of local ones, reducing the likelihood of co-present sociability. However, as we have seen in earlier chapters, since the growth of big metropolises people have often avoided interacting with strangers in public. In fact, they have had to develop mechanisms to manage these interactions and to protect themselves from the perceived "chaos" of public spaces. The mobile phone works as a more recent interface individuals use to manage their interactions with public spaces.

Puro was not alone in his claims. Gergen (2002) wrote that when using a mobile phone, "we are present but simultaneously rendered absent; we have been erased by an absent presence . . . One is physically present but is absorbed by a technologically mediated world of elsewhere" (p. 227). Even Howard Rheingold

(2002), forceful advocate of the potential of new technologies to create communities, observed that on trains and buses in Japan passengers preferred to text somebody who was physically absent than talk to other people in the same vehicle. However, as Wolfgang Schivelbusch (1986) noted, since the development of the train, when passengers were "forced" to spend long hours facing each other in a U-shaped wagon, people cultivated the habit of reading on trains because they were trying to avoid the awkward situation of talking to a stranger who was sitting nearby. Mobile phones work similarly. Among other functions, they also help people to filter out awkward situations in public spaces. For example, frequently people pretend they are talking on the phone in order to feel safer in a public space, or to avoid interacting with people nearby (Caronia & Caron, 2004). There is even a *Facebook* page titled "Pretending to talk on your cell phone when you see a stranger"[1] that exemplifies this common use of the mobile phone.

Because mobile phones also allow people to talk to remote others, these technologies have often been blamed for not only removing people from their surroundings, but also for privatizing public spaces (Höflich, 2006; Puro, 2002). Comparing mobile phones to landline telephones, De Gournay (2002) observed that "people are talking outside the home, so that the public is put into the position of a 'voyeur,' involved whether it likes it or not in the secrets of households or couples, accidentally overhearing private conversations in public places" (p. 108). The voyeuristic nature of the mobile phone call is one of the reasons why mobile phone use in restaurants and public transportation is perceived to be particularly annoying (Ling, 2004; Nickerson, Isaac, & Mak, 2008; Okabe & Ito, 2005). Of course, overhearing private conversation in public did not start with mobile phones, but mobile phone conversations are more noticeable and perceived as more intrusive than face-to-face conversations. This perception has been related to at least three facts: the novelty of the technology, the loudness that people speak on their mobile phones (due to the lack of feedback noise), and, most importantly, the fact that people can only hear one side of the conversation (Monk, Carroll, Parker, & Blythe, 2004; Monk, Fellas, & Ley, 2004).

Joachim Höflich (2006) goes further and suggests that this privatization of public spaces is also shown by the fact that social networks sustained by mobile phones are generally composed of a small number of people who already know each other and probably already have strong personal ties. For him, rather than helping people forge new heterogeneous social connections, the mobile phone emphasizes comfort, strong ties, and homogeneity. These social connections, according to Ichiyo Habuchi (2005), create "a zone of intimacy in which people can continuously maintain their relationships with others who they have already encountered without being restricted by geography and time" (p. 167). That is part of the reason why mobile phones were said to "free" users from physical space (Geser, 2004).

The ability to overcome geographical distance has led some mobilities scholars to categorize the mobile phone as a form of virtual mobility (Kellerman, 2006;

Urry, 2000). The idea of virtual mobilities emphasizes the movement and flows of information and messages across long distances. Although these flows are supported by physical infrastructures, the connection to nearby people and places is de-emphasized. According to this logic, the mobile phone is viewed as a device that enables remote communication and the formation of close-knit social networks while on the move, at the expense of local connections. These types of remote social networking in public space have been called telecocooning (Habuchi, 2005), selective sociality (Matsuda, 2005), and networked individualism (Wellman, 2002). Developing the idea of networked individualism, Barry Wellman suggested that people who use mobile phones are not necessarily anti-social, but they are social in a very peculiar way: They are socially connected, but "people's awareness and behavior are in private cyberspace, even though their bodies are in public space" (2002, p. 6). In a study about the use of wireless devices (including mobile phones) in public plazas, Hampton et al. (2010) observed that while wireless internet users are actively interacting with their remote social networks, they often physically withdraw from the public space by sitting at a private corner, and therefore pay reduced attention to their surroundings. As a result, because they look less approachable, they have less opportunity to interact with others around them (Hampton & Gupta, 2008). As such, both Wellman and Hampton et al. emphasize how mobile technologies disconnect users from their surroundings by allowing distant connections at the expense of nearby ones.

While these scholarly discussions have all been useful for understanding the relationship between mobile phones and public places, they tend to lack the nuance of a more micro-level understanding of mobile phone use. This perceived disruption of public spaces depends on how users and non-users deal with the presence of the technology. People can choose to see remote connections as a disruption of the co-present situation, or they can choose to understand these connections as *part of* the public space. In her extensive study of early mobile phone use around the world, Sadie Plant (2001) described different ways co-present groups of friends dealt with incoming mobile calls. The "innies" would leave the group as to not disturb the ongoing conversation. The "outies" would stay in the group, and often engage in both activities at the same time. While both "innies" and "outies" would not perceive an incoming call as disrupting the co-present situation, Plant noted that there was yet a third group to which a phone call would be understood as annoying and as interrupting the co-present social situation. Mobile phone users in these groups would then face the disapproval of others and often answer the call by turning their backs on co-present people, further emphasizing the disconnection from the present situation. The value of Plant's early observations is that it focuses on the co-present social interactions, rather than on the technology that disrupts it.

Another useful approach to the social use of mobile phones in public spaces is Scott Campbell and Nojin Kwak's (2010) recent study on how mobile phone users do engage with others in public. Contrary to their initial hypothesis that the use of mobile phones prevents interactions with strangers, they found that, in some situations, mobile phone use does increase such social connections. It depends, as they emphasize, on how people use the technology. For example, people who used their mobile phones to access news and information tended to share some of the contents they read with others nearby, even if these others were strangers.[2] Similarly, when using mobile phones for micro-coordination (that is, quick calls among family members and close friends to schedule and re-schedule everyday tasks), they also observed an increase in interaction with strangers. However, they believe this is simply because these users are in public, rather than coordinating from a private location, such as home or office. Finally, among people who mostly used their mobile phones for relational use, that is, to have conversations to close friends and family, there was no evidence of interactions with strangers. Campbell and Kwak's study shows that it is too simple to make blanket statements about mobile phones disconnecting people from their surroundings. Instead, the way people use the technology is varied and depends on the social context and dynamics in which they are used.

However, the mobile phone literature features a number of sweeping statements about mobile phone use in public space. Puro (2002), for example, believed that "public space is doubly privatized because mobile phone users sequester themselves non-verbally and then fill the air with private matters" (p. 23). Nevertheless, as we discussed in the previous chapter, such discourses about the privatization of public spaces assume that the concepts of public and private are static and clearly defined (Sheller & Urry, 2003). Claiming that the mobile phone damages good public spaces imagines that any private activity that distracts an individual's attention from the co-present situation is damaging to the construction of public spaces. This view fails to account for the ongoing permeability and fluidity between private and public we discussed in the first two chapters, instead viewing the public as something fixed that is weakened by remote connections.

Some scholars do subscribe to a more nuanced view on the shifts between private and public, suggesting that what we are witnessing with mobile phone use is not simply a privatization of public spaces, but rather a subtler and more fluid redrawing of borders between public and private (Goggin, 2006). This constant negotiation of social norms and of the borders between the private and the public is addressed by Rich Ling (2008) in his book *New Tech, New Ties*. Drawing from Goffman, Ling suggests,

> The pose of using a mobile phone is a request for civil inattention. Indeed, the ringing of the telephone and our disappearance into the sphere of mediated interaction while holding the telephone up to one ear can be used to generally remove us from the demands of the local scene. There may even be the sense that we temporarily colonize a portion of the public

sphere when talking on a mobile phone. However, this can be an ongoing negotiation with others who are present and who also have a legitimate reason for using the space.

(p. 106)

Like Ling, Lee Humphreys (2005) analyzed mobile phone use in public space as a process of ongoing negotiation. As such, she rejects the idea of a complete privatization of public spaces, arguing, "People still engage on some level within their participation unit even if it is not active or direct interaction. Despite people having private conversations in public spaces, they are still part of a larger interaction group" (p. 371). As happened with the book, the iPod, and the Walkman, talking on a mobile phone "helps people maintain their minimal main involvement" with their physical surroundings" (p. 369).

Although these studies do emphasize mobile phone users' awareness of their surrounding space, they focused primarily on how mobile phones help users manage the relationships with co-present people, rather than managing relationships between people and the places they move through. With the goal of focusing on the relationships between users and places, Weilenmann (2003) recorded mobile phone conversations among teenagers to find out how people articulate location while on a phone call. She found that people more frequently ask, "What are you doing?" rather than "Where are you?" showing that the question of location was generally subordinated to the coordination of activities. Similarly, in his article "Why do people say where they are during mobile phone calls," Eric Laurier (2001) discussed how the articulation of place provides context for mobile phone conversations. Among other things, he acknowledges that people talk about where they are in a phone conversation mostly in order to contextualize the dialogue to the person at the other end of the line, and this contextualization generally serves the purpose of coordination with remote others, rather than interacting with people and things nearby. As he explains, "For Penelope and Sharon to co-ordinate their affairs then, some sense of shared context has to be accomplished by them" (p. 11).

From the examples above, we see that mobile phones have been primarily studied for their ability to coordinate and manage time and relationships with others, often independent of physical location (De Gournay, 2002; Licoppe & Heurtin, 2001; Ling, 2004; Ling & Yttri, 1999, 2002; Mante, 2002). Rich Ling and Birgitte Yttri (1999) named this ongoing series of short phone calls between individuals who share strong ties *micro-coordination*. In micro-coordination, space is generally analyzed as subordinated to time, that is, people often talk about the place where they are in order to schedule and re-schedule mundane daily activities. What's more, most of these studies suggest that mobile phone users need to articulate place because "mobile phones bring a loss of the tie between place and member" (Laurier, 2001, p. 17). Here again we see a disconnection between mobile phone users and places.

Very few scholars focused on how mobile phones connect users to places and how they work as a filter to public spaces, helping users manage their interactions with their surrounding space. One such example is the work of Christian Licoppe and Jean-Philippe Heurtin (2002). They suggest,

> The mobile phone's use as a key resource for successful coordination over space and time lies in part in its strong *impact on perceptions of space*. On the one hand, it remodels the perception of the ambient space for its user. On the other hand . . . the person who calls or is being called by the mobile phone user can no longer assign a definite location to the other person from either the geographical or the social perspective.
>
> (p. 96, emphasis added)

Licoppe and Heurtin were interested in exploring how using a mobile phone remodels the perception of place for its user. As we will see in the next section, it is no wonder that Licoppe's subsequent research moved into the analysis of location-based technologies because these technologies literally help people find information and people from the space around them, making the connection with places (and locations) explicit rather than implicit. However, Licoppe and Heurtin also acknowledge that there is a loss in the *sense of place* (Meyrowitz, 1985) with which mobile phone users need to negotiate.

When studying mobile phones as voice and text-communication devices, it is hard to depart from the focus on remote interaction and from the idea that mobile phones interrupt co-present situations. It is true that the ability to communicate while moving through public spaces implies a qualitative change in the process of communication. But this situation changes with location-aware mobile technologies.

Location Awareness

When D. H. Ring at Bell Labs developed the cellular concept in 1947, he proposed that mobile phones should exchange radio signals with the three closest cell towers (Agar, 2004; Farley, 2005). By triangulating waves, mobile phones can not only transmit voice, but also be located. The ability to locate mobile phones was what the U.S. Federal Communications Commission (FCC) took advantage of when, in 1997, it required that all mobile phones in the United States be locatable for emergency calls, that is, if somebody called 911 from their mobile, their wireless carrier should provide emergency services with the mobile phone's number and the cell site or base station transmitting the call (Goggin, 2006, p. 196). In 2001, with E911 phase II, the U.S. government required that the location of the mobile device be provided to emergency services with an accuracy of 50–100 meters. As we will see next chapter, some privacy advocates were concerned about such laws, but the mobile phone's location-aware capabilities did not become widely available to the public in the United States until the popularization of 3G phones in 2008.

Japan, however, has long been at the forefront of the development and adoption of new mobile services and applications. The first commercial analog mobile phone service was released in Tokyo in 1979 (Farley, 2005; Goggin, 2006). The i-mode service (the Japanese mobile standard for accessing the Internet) was launched by NTT DoCoMo in 1999 (de Souza e Silva, 2006), and the Japanese were therefore one of the first people in the world to use their mobile phones for more than voice and text messages. LBS have also been available in Japan since that time. The first types of LBS popular in Japan were real-time mapping and direction applications. As early as 2001, the Japanese were already using i-mode-based LBS that allowed users to find information about the closest restaurant and shops. The *i-area* service, for example, delivered location-specific weather, traffic, and dining information to users all over Japan (DoCoMo, 2002).

Scandinavian countries also pioneered the development of LBS. In 2001, Swedish start-up company It's Alive launched the world's first commercial location-based mobile game (LBMG): *Botfighters*. LBMGs take advantage of players' locations to transform the experience of mobility through physical space into game play. LBMGs are generally multiplayer and as such allow players to interact with each other depending on their relative distance in physical space (de Souza e Silva, 2009; de Souza e Silva & Sutko, 2009). Some LBMGs also require users to interact with digital information "attached" to locations by, for example, "collecting" digital objects with the mobile phone. *Botfighters'* design was based on traditional first-person shooter games. The game's goal was to virtually "shoot" other players via text messages. In order to play, one had to go to the game website and create a robot that became the player's avatar. The robot was armed with guns and shields and downloaded to the player's mobile phone. Whenever another robot (player) was within a specific radius, she would get a text message that said "robot 'X' nearby" (Figure 3.1). She could then send another text message back to the server to shoot the enemy robot. Successful shots depended on the distance between players and on the types of guns one possessed. For example, a sniper was able to kill another robot up to one kilometer away (de Souza e Silva, 2009; Sotamaa, 2002).

Botfighters' shots, however, were not very precise because the game was developed based on the mobile phone's cellular triangulation capability. Location-awareness via GPS is much more accurate, but at the time of the game's development, that technology was still not widely available. That changed in 2000 when the Clinton administration removed the signal degradation from civilian GPS devices, allowing people and companies to acquire precise enough signal to find specific objects and locations via latitude and longitude (lat/long) coordinates.

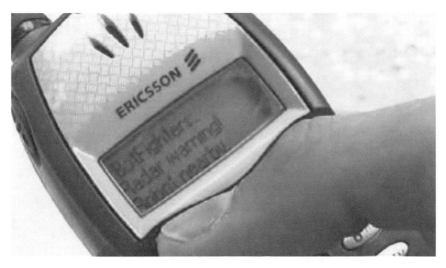

Figure 3.1 The *Botfighters'* mobile interface from its promotional video. Copyright: Tom Söderlund.

Early GPS Experiments: Mapping and Games

One of the early uses of GPS devices was for entertainment. *Geocaching* was one of the first GPS treasure hunt types of game (Figure 3.2). To play, people first used the game's website to search for hidden caches and downloaded the caches' lat/long coordinates to their GPS. They could then precisely locate the cache, documenting their discovery back on the website (de Souza e Silva & Hjorth, 2009; Gordon, 2009).

GPS was also widely embraced by early locative media artists (Hemment, 2004, 2006; Tuters & Varnelis, 2006). As a mix of LBMG and artistic performance, Blast Theory and the Mixed Reality Lab's *Can You See Me Now?* (2001) was one of the first innovative uses of GPS technology to create a game experience that took place simultaneously in physical and digital spaces. *Can You See Me Now?* was a type of chase game that took place in hybrid spaces (Figure 3.3). In the game, street runners had to catch online players. While online players moved through a digital model of the city, street runners moved around the physical city and could visualize the relative position of online players on a personal digital assistant (PDA) attached to a GPS device (de Souza e Silva, 2009; Flintham et al., 2003; Goggin, 2006; Tuters & Varnelis, 2006).

Other early locative media art projects focused primarily on the mapping and tracing capabilities of the GPS technology. For example, artist Jeremy Wood recorded several of his movements through space in a series of projects called *GPS Drawing* (2000). Taking Wood's idea further, Ester Polak and the Waag

Figure 3.2 A Geocache container. Copyright: John Robb.

Figure 3.3 Can You See Me Now? street runner and PDA mobile interface. *Can You See Me Now?* is a collaboration between Blast Theory and the Mixed Reality Lab, University of Nottingham (2001–2011).

Society's *Amsterdam Real Time* installation (2002) tracked participants in real time, and transmitted a live representation of their paths through downtown to the exhibition *Maps of Amsterdam 1866–2000*. In doing so, the project forced visitors to contrast the traditional maps of the city to the live constructed maps of the urban space created by the mobility of people (Figure 3.4). Such projects and games, however, used GPS receivers as stand-alone devices. Soon GPS technology was incorporated into mobile phones, allowing the development of social interactions that took advantage of people's location in physical space.

In March 2003, NTT DoCoMo announced the first GPS mobile phone in Japan (DoCoMo, 2003). In the same year, *Mogi*, the first commercial LBMG that took advantage of GPS-phone functionality was launched in Tokyo. *Mogi* was developed by French start-up company Newt Games and commercially

Figure 3.4 The *Amsterdam Real Time* interface shows GPS traces of people who walk around the city of Amsterdam. Copyright Ester Polak/Waag Society (2002).

released by Japanese provider KDDI (Licoppe & Guillot, 2006; Licoppe & Inada, 2006, 2009). Similar to *Geocaching*, the game was based on a treasure hunt, with the difference being that now players had to look for digital (rather than physical) objects and creatures that were "attached" to specific locations with lat/long coordinates (Figure 3.5). Whenever a player was less than 400 meters from the object, she could capture it to complete a collection. Some digital creatures, however, would only appear on the game's radar (on the mobile phone screen)

Figure 3.5 The *Mogi* mobile client interface displaying nearby players in the city of Tokyo. Reproduced by permission of Newt Games.

at specific times of the day, requiring players to organize what Christian Licoppe and Yoriko Inada (2006) called "expeditions" to specific parts of the city at unusual hours to capture them. Other players reported changing their route to work in order to play the game and capture more objects that were available on a bus ride but not on the subway (Licoppe & Inada, 2006). *Mogi* was then followed by other LBMGs such as *Alien Revolt* in Brazil (de Souza e Silva, 2008), *Swordfish* and *Torpedo Bay* in the USA (Xiong, Ratan, & Williams, 2009), and the J2ME[3] version of *Botfighters* in Sweden (de Souza e Silva & Hjorth, 2009).

Besides experimental LBMGs, locative media artists also pioneered the development of applications for GPS-enabled mobile phones. In 2004, Blast Theory and the Mixed Reality Lab premiered *I Like Frank*, the world's first 3G hybrid reality game in Adelaide, Australia (Figure 3.6). *I Like Frank's* (2004) participants unfolded a narrative similar to their previous piece *Uncle Roy All Around You* (2003), except players were equipped with GPS-enabled phones. Instead of self-reported position as in *Uncle Roy*, *I Like Frank* players had their location automatically tracked and shared with online players on a 3D model of Adelaide. Street players had 60 minutes to find Frank, and to accomplish their mission, they had to pair with online players and share place-specific information (de Souza e Silva & Sutko, 2008).

LBMGs have also shown potential as education tools (de Souza e Silva & Delacruz, 2006; Delacruz, Chung, & Baker, 2009). One of the first educational LBMGs to take advantage of a GPS phone was *Frequency 1550*. Developed by the Waag Society and educators at the University of Amsterdam, the game was piloted during three days in February 2005 (Admiraal, Akkerman, Huizenga, & Zeijts, 2009; de Souza e Silva & Delacruz, 2006). *Frequency 1550's* goal was to teach 11–14-year-old students the history of medieval Amsterdam. Students were split into two categories: merchants and beggars. Each category was then divided into two teams: One team went on the streets of Amsterdam equipped with GPS-enabled 3G mobile phones and the other team stayed at a remote classroom

Figure 3.6 The *I Like Frank* mobile client interface showing a map of the city of Adelaide. *I Like Frank* is a collaboration between Blast Theory and the Mixed Reality Lab, University of Nottingham (2003).

location where they could help, guide, and interact with the students on the streets. Groups combining online and street players needed to solve location-based assignments embedded into the game narrative, which revolved around gaining citizenship in the city of Amsterdam. One of the goals of *Frequency 1550* was to situate the learning experience in the physical space. Because students interfaced with physical locations and with each other through GPS-enabled mobile technologies, the game designers focused on situated learning approaches that emphasized context. For example, one of the assignments of the game intended to trigger environmental awareness required city team students to walk the actual route of a procession as a pilgrim, film the route, and send the video to the headquarters students, who would in turn select on a map the parts of the medieval city where the pilgrims walked (Admiraal et al., 2009).

Games and play have been recognized as important activities that motivate students to learn (Barab, Sadler, Heiselt, Hickey, & Zuiker, 2007; Gee, 2007; Kafai, 2006; Vygotsky, 1978). Many scholars and educators relied on the theorized benefits of situated and social learning to build virtual worlds that immersed students into real life-like situations (Barab, Thomas, Dodge, Carteaux, & Tuzun, 2005; Dede, Nelson, Ketelhut, Clarke, & Bowman, 2004). LBMGs, however, actually situate the learning experience in physical locations. Rather than simulating a real-life situation in a virtual environment, students have the opportunity to interact with real people and information that is attached to locations (de Souza e Silva & Delacruz, 2006; Delacruz et al., 2009). While mobile phones have often been criticized in classrooms and schools because of the way they distract students (Campbell, 2006; Katz, 2005), LBMGs show how mobile phones can contribute to education by connecting people to locations. As we have seen in the introduction, locations acquire new dynamic meaning due to the always changing amount of location-based information that is embedded in them.

Mobile Annotation: The City as a Canvas

Besides gaming and mapping, mobile annotation projects also contributed to the construction of locations. Although very different from each other, these projects had a common characteristic of employing GPS-enabled phones as interfaces that allowed users to "attach" information to places and to retrieve location-based information. For example, the UK-based art group Proboscis developed the project *Urban Tapestries* (2002–2004) to investigate what they called "public authoring." The project encouraged participants to use their mobile phones to include geographical coordinates in the metadata of stories, pictures, sounds, and video and "attach" them to specific locations, embedding social knowledge into the fabric of the city. This digital information could then be retrieved when another user was near the location of the information. *Urban Tapestries* enabled participants to become authors and co-creators of their hybrid environments (Figure 3.7).

Figure 3.7 The *Urban Tapestries* mobile interface. The map on the PDA shows where location-based information can be found in the city, and how each piece of information is linked to other information, forming threads. Copyright Proboscis 2004. Photo John Richard at Proboscis (2004).

Around the same time, research and development departments in companies such as Siemens and HP were also thinking about mobile annotation. In 2005 Siemens, in cooperation with researchers at the University of Linz in Austria and the Ars Electronica Center, announced the development of a system called *Digital Graffiti* by which a user would be able to send a text message to any geographic location by assigning lat/long coordinates to it. With *Digital Graffiti*, if a user were to enter the target text message location, the message was then displayed on her screen. Two possible user groups for this technology envisioned by Siemens were outdoor tourists, who would be able to read information about monuments when approaching specific places, and location-based advertisers, who would send promotional coupons to people nearby who subscribed to the "advertising mode."

Also in 2005, HP released an authoring toolkit called *Mscape*, which enabled users to "attach" information to locations in the form of text, sounds, interactive games, walks, and textures. To create a "mediascape," users dragged sounds, text, and images on a map of any landscape using HP's special software. Their newly created mediascapes could then be downloaded to mobile phones. People could experience the mediascapes when approaching the location of the digital information. An example of an interactive *Mscape* game took place in the Tower of London. The game enabled users to re-enact historical moments and learn about the history of the tower. Once inside the tower, users loaded up the mediascape and listened to the instructions. To start one of the missions, users had to tap on the picture of one of the prisoners. Then, if asked to go to a specific location, the cursor button on their device would bring up a map, which showed their location and where they needed to go (Hewlett-Packard, 2005).

Projects such as *HPMscape* and to some degree *Urban Tapestries* encouraged users to follow a location-based narrative to participate in a game or experience a museum space; they can be traced back to earlier audio-walk projects, as we will see in Chapter 6. Such projects have been highly popular not only among research and development units and locative media artists (Knowlton, Spellman, & Hight, 2002; Rueb, 2005), but also among museums that wanted to create a more personalized visitor experience (Sparacino, 2002). Together with the mapping and gaming projects described in the former section, they contributed to highlight the changing meaning and importance of locations.

The Commercial Development of Location-Based Services

Most of the experiences described in the former section, although crucial for the development of today's LBS, were created in the domain of art and research and did not reach a large user base. Throughout the 2000s, the number of commercially available 3G phones equipped with GPS slowly increased, but it was not until 2007, with the release of the original iPhone, that the development of location-based applications left the experimental domain of art and research and became more accessible to the general public. The causes of this sudden increase in location-based apps can find a parallel in the popularization of the i-mode in Japan. As mentioned earlier, i-mode became a popular mobile platform because it was easy for users to create applications for the platform. When 3G phones were first available, mobile phone providers invested great sums of money in advertising for the new platform, focusing on the opportunity to stream videos and watch TV on their phones (Groening, 2010; Wilson, 2006). But contrary to their expectations, early 3G phones were slow to take off, and it took years for the mobile carriers to begin to make back the money they had invested in 3G networks (Goggin, 2006; Wilson, 2006). According to Jason Wilson (2006), one of the reasons for this initial failure was a lack of vision from mobile providers that insisted in paying their debt by charging users too much for accessing mobile applications. He contrasts this top-down approach to the development of the so-called web 2.0. The World Wide Web (WWW) experienced enormous growth in the first decade of the twenty-first century partly because of the development of a diverse set of platforms (such as blogs, wikis, and social networking sites) that allowed users to create and upload content.

3G mobile phones have just begun to follow a similar trajectory, as can be seen by the growing importance of applications (apps). As of October 2011, the iPhone app store now features over 300,000 applications, and Google's Android app marketplace has over 100,000 applications. App development is no longer only in the hands of mobile phone providers, so users can develop content for the platform. Furthermore, there are apps that work as authoring tools to help other users with fewer programming skills to create other location-based experiences, such as SCVGR[4] and 7*Scenes*. Because users are now able to create content for mobile platforms, apps have been critical for the development and popularity of LBS. Many of the most popular apps today feature location-aware capabilities.

Drawing from the experimental development of location-based applications, we can observe that many of the early commercial apps involve some kind of digital annotation, social networking, or mapping—or a combination of the three. Location-based advertising (LBA) is another quickly growing area, which we deal with in more detail in the following chapter. Digital annotation apps, such as *WikiMe*, *Geopedia*, and *Yelp* allow users to upload and visualize nearby location-specific information on their mobile screen. With *WikiMe*, for example, users can search for *Wikipedia* articles with their mobile phone. However, instead of retrieving results by topic, as happens with a regular search, *WikiMe* users retrieve articles relevant to their location (de Souza e Silva & Frith, 2010; de Souza e Silva & Sutko, 2011). If somebody is standing close to the Eiffel Tower in Paris when loading *WikiMe*, she will be able to see all the *Wikipedia* articles written about the area surrounding the Eiffel Tower on her screen. *Geopedia* originally worked very similarly to *WikiMe*, but its latest version also allows users to access nearby *Flickr* photos as well.

Many location-based apps enable users to actively contribute to the information landscape of locations by, for example, writing a *Wikipedia* article, uploading a *Flickr* photo with lat/long coordinates, or (as is the case with *Yelp*) writing a restaurant review. Other types of mobile annotation applications display local news stories, discussion threads, and location-based blogs relevant to the user's physical location. For example, the *Radar* service of Outside.in lets iPhone users switch among different scales to find news within 1,000 feet of their current location or stories from the surrounding neighborhood. As with *WikiMe*, users can choose between a list and a map interface.

Some newer mobile annotation applications include augmented reality (AR) elements, as is the case with *Wikitude*, *ZipReality Real State*, *Layar*, and *TwittARound*. Although AR platforms have been envisioned as "killer-apps" in research labs for quite a long time (Feiner, MacIntyre, Hollerer, & Webster, 1997; Mooser, Wang, You, & Neumann, 2007), they did not become popular until the release of smartphones capable of running AR applications because of their improved cameras and GPS. *Layar* was one of the first AR apps. It allows people to hold up their phones and visualize "layers" of information retrieved from other location-based services, such as information about dining, entertainment, and dating

(Figure 3.8). The new *Layar Vision*, released as a beta app in 2011, allows users to scan specific objects and images in the physical world and retrieve information about them. So, for example, if a *Layar Vision* user scans the poster of a music festival that is hanging on the wall of a bar, the application will automatically recognize the object and asks the user if he would like to buy tickets to the show.

However, to date, most user–interaction with AR information is restricted to application developers building specific apps to display specific content. For example, *Yelp* built a specific AR app to display restaurant reviews; other apps such as *TwittARound* display only location-based tweets. As an alternative to this scenario, the College of Computing at Georgia Tech developed the first augmented reality browser, which they called *Argon*. *Argon* is an open standard platform that allows people to create AR content using existing web development toolsets. The idea is that any individual or business could be able to create an AR channel on their already existing websites, which would then be displayed through the AR app. The development and availability of applications such as *Argon* suggest that the interactions with physical spaces will likely include more AR elements—users will not only be able to read location-based content about the environment around them, but also to visualize location-based information as layers superimposed on this environment.

So far we have been describing location-based apps focusing on their mobile annotation and mapping capabilites. But increasingly some LBS also include a social networking element. These apps not only allow users to access location-

Figure 3.8 The *Layar* interface shows digital information superimposed onto the background of the city. Image reproduced with permission from Layar.

specific information from their environment, but also to socialize with other users who are nearby. As we have seen, the first types of location-based social networks (LBSNs) were LBMGs, such as *Botfighters* and *Mogi*. Around 2008, however, partly influenced by the popularity of social networking sites such as *Facebook*, some of the early LBSN apps for the iPhone enabled users to interact with each other depending on their location, but without the addition of killer robots or digital objects. *Loopt* was one of the first of this kind. The original version of *Loopt* allowed users to add friends to their social network and visualize their location on a map on their mobile phone screen. Users could post status updates that would automatically insert the location of the post. While *Loopt* just displayed the location of people who had previously accepted a friendship request, the *LooptMix* feature enabled users to visualize the location of any other *LooptMix* user in the vicinity (de Souza e Silva & Frith, 2010).

Other LBSN apps that developed around the same time were *Brightkite*, *Whrll*, and *Centrll*. *Brightkite* worked similarly to *LooptMix* because it did not restrict the visualization of location by friendship, but rather by distance. *Brightkite* users could choose to see the location of any other *Brightkite* user within a block (200 meters), within a neighborhood (2 km), within an area (4 km), within a city (10 km), or within a region (100 km).

While *Brightkite* and *Loopt* focused on location, *Whrll* emphasized the storytelling potential of such applications by encouraging users to tell stories about their location, a characteristic that is also a legacy from the first location-based audio-walks, as we will discuss in Chapter 6. As of 2011, there are a myriad of LBSNs apps, the most prominent ones being *Latitude*, *Loopt*, *Gowalla*, and *Foursquare*. Interestingly, *Foursquare* became popular partly because it included the gaming component, a characteristic originally present in LBMGs.

Released in 2009, *Foursquare* was developed by Dennis Crowley and Naveen Selvadurai. Nine years earlier, when Crowley was a student at New York University, he created the mobile service *Dodgeball*. *Dodgeball* was a mobile social network, but not location-based per se because, like Blast Theory's *Uncle Roy All Around You,* it used self-reported position by requiring users to send a text message to the server to say, "I'm here." *Foursquare* expanded the social networking features of *Dodgeball* by taking advantage of the location-aware capabilities of the mobile phone. In addition to the social networking features of other LBSNs such as *Loopt*, users also gain points and badges by checking into places. If somebody checks in repeatedly at the same place, she eventually becomes the "mayor" of that location. Mayors can have multiple privileges, including receiving discount coupons and free treats from the venue they "control." Like in *Brightkite*, *Foursquare* users are also able to see the location of others checked in nearby, including the mayors. By clicking on another user's profile, depending on their privacy settings, they can also access their location history. Following *Foursquare*, other LBSNs also added gaming elements. These ludic elements have been

Figure 3.9 The *CitySense* interface. Reproduced by permission of Sense Networks, Inc.

incorporated into LBSNs to foster connections between users and to establish a context for connecting with locations. For example, *Whrrl* introduced "badges" and "societies" as tools for motivating participants, and *Loopt* launched what it called *Loopt Star*, an app that gives users rewards such as free tickets and food depending on where they check in.

Other early types of LBSNs do not attempt to identify individual users, but instead show mobility and activity patterns through the city. These are what Daniel M. Sutko and Adriana de Souza e Silva (2011) called anonymous LBSNs. For example, *CitySense* was released in 2008 as the "first mobile application to take millions of data points to analyze aggregate human behavior and to develop a live map of city" (CitySense, 2008). Based on Sense Networks platform, the application draws from available GPS points and Wi-Fi positioning data to show concentrations of people as a heat map over the city of San Francisco. More popular places are shown as darker red clouds. *CitySense* is marketed as a "real time night-life application that shows where the action is" because *CitySense* users open the application not to interact with individual users, but to connect directly to places (Figure 3.9).

Constructing the Meaning of Locations

New hybrids of LBSNs/LBMGs, such as *Foursquare* and *Loopt*, are becoming increasingly popular,[5] but the types of social and spatial interactivity they enable are still very limited. In 2007, Dourish, Anderson and Nafus claimed that "[the design of] location-based systems fail to acknowledge the lived practice of urban life" (p. 2). Although Dourish et al. wrote before the development of most current apps and "smartphones" such as the iPhone, they raise an important point. As we will see in Chapter 6, location-aware technologies do have the potential to enable users to navigate cities in new and unprecedented ways, in ways that go beyond going from place to place, or simply checking in specific locations. However, as we will see in Chapter 5, the current development of most LBS is strongly grounded upon location-based advertising as a source of revenue, which forcefully affects how users may interact with their surrounding space and other people in it.

As we discussed in the last section, since their inception commercial LBMGs, LBSNs, and LBS in general were designed to help users "attach" information to specific locations, find other people in the vicinity, and find directions. The problem with this approach, according to Dourish et al. (2007), is that they often frame mobility as a problem to be solved (rather than the practice of movement itself). The designers of technology often focused on movement from point A to point B, but often failed to take the meaning of users' environment into consideration, that is, the in-between space. Still today, the most popular LBS are navigation services and services that aim at helping people find other people and information. This is not intrinsically a problem. However, when the types of information and people one finds through LBS are affected by location-based advertising, it does raise a question about the meaning of locations (and consequently public spaces) that is being constructed through location-aware technology use.

The early locative media artworks exemplified in the former section framed mobility as a new opportunity for interacting with people and locations. Although we saw that the development of commercial LBS owes a great deal to early experimental media artists and researchers, these services still must develop further if their goal is to transform urban mobility and the interaction with locations into a more meaningful experience. We can see some hints of this development in newer applications. For example, *CitySense* emphasizes movement through urban spaces as collective patterns, rather than individual experiences. And *Foursquare* focuses on connecting users and motivating them to connect to locations on their own way, rather than just finding a final destination.

Dourish et al. (2007) stated that the development and design of urban computing "shape, are shaped by, and mediate our experience of urban space" (p. 1). Therefore, instead of focusing on how mobile technologies contribute to withdrawing people from their surroundings, as was generally the case with mobile phones and Walkmans, the study of location-aware mobile technologies urges us to consider how they shape and influence our interactions with locations and public spaces. Looking back at this brief history of location-aware technologies, we see the potential they have to not only construct locations, but also to connect users to the people and information in those locations. In the beginning of this chapter, we discussed how mobile communication scholars focused on how mobile telephony helped users forge remote connections with strong ties to social networks at the expense of co-present social interactions and connections to nearby places. Like the other interfaces previously analyzed in this book, mobile phones were claimed to bring private activities into the public and thus privatize public spaces. The mobile phone supposedly did so by bringing personal conversations into a space of heterogeneity shared by strangers. Frequently, these conversations with remote others were perceived not only as a privatization of space, but also as disconnecting users from their surrounding space. Mobile phones, however, work similarly to other mobile technologies: They enable users to selectively interact with their surroundings. By texting somebody while on a bus, or by talking to a friend while walking on the streets, mobile phone users alternate their attention between the remote and the co-present situation.

Although early mobile phones enabled users to manage their interactions with their surrounding space, location-aware phones make this connection with the local environment even stronger. By explicitly enabling people to interact with information that is location-specific, the use of these technologies in public spaces forces us to look differently at locations and the spaces we move through. We can now interact with different types of location-based information, which, as we will discuss in Chapter 6, leads to new ways locations are presented because physical space is merged with digital information.

While certain location-based services, such as navigation and mapping services, have seen fairly rapid adoption, the number of people using location-based social networks is still relatively small. According to a survey by the Pew Research Center's Internet & American Life project conducted between August and September 2010, only 4 percent of adults in the United States who access the Internet with their mobile phone use LBSNs (Sickhur & Smith, 2010). Among this group, the majority is between 18 and 29 years old, which is a fairly typical early adopter group. According to diffusion and adoption theory (Rogers, 1995), early technology adopters are generally young. However, even at this early adoption stage, the emergence of LBS, and especially LBSNs, has already raised passionate arguments about potential social implications, including invasions of locational privacy, the possible dangers of stalking, and differential power issues in social relations. In the following chapters, we will address these implications in more detail, discussing them in the context of privacy (Chapter 4), power (Chapter 5), and the presentation of location (Chapter 6). All of these issues are framed within a discourse of control; in other words, the meaning of privacy, power, and public spaces is tied to how people control digital information that is attached to locations. In addition, location-aware mobile technologies interface in new ways the relationship between people and places by foregrounding the importance of locations in our social and networked interactions.

Notes

1   http://www.facebook.com/pages/Pretending-to-talk-on-your-cell-phone-when-you-see-a-stranger/451497485262.
2   Note here that in this situation the mobile phone is used as more than a two-way communication voice device, and its use is therefore different from what was studied in earlier literature.
3   J2ME stands for "Java platform, micro edition." When mobile phones started to run Java, they could handle simple graphics rather than just text.
4   SCVGR is a location-based game authoring platform that allows common users to create their own games and location-based experiences. We will discuss it more in depth in Chapter 6.
5   As of June 2011, *Foursquare* announced it reached 10 million users worldwide (Foursquare, 2011; Swartz, 2011).

References

Admiraal, W., Akkerman, S., Huizenga, J., & Zeijts, H. V. (2009). Location-based technology and game-based learning in secondary education: Learning about Medieval Amsterdam. In A. de Souza e Silva, & D. Sutko (Eds.), *Digital cityscapes: Merging digital and urban playspaces* (pp. 302–320). New York: Peter Lang.

Agar, J. (2004). *Constant touch: A global history of the mobile phone.* Cambridge, UK: Icon.

Barab, S., Sadler, T. D., Heiselt, C., Hickey, D., & Zuiker, S. (2007). Relating narrative, inquiry, and inscriptions: Supporting consequential play. *Journal of Science Education and Technology, 16*(1), 59.

Barab, S., Thomas, M., Dodge, T., Carteaux, R., & Tuzun, H. (2005). Making learning fun: Quest Atlantis, a game without guns. *Educational Technology Research and Development, 53*(1), 86–107.

Blast Theory, & Mixed Reality Lab (Artist). (2001). *Can you see me now?* (Hybrid reality game).

Blast Theory, & Mixed Reality Lab (Artist). (2003). *Uncle Roy all around you* (Hybrid reality game).

Blast Theory, & Mixed Reality Lab (Artist). (2004). *I like Frank* (Hybrid reality game).

Campbell, S. (2006). Perceptions of mobile phones in college classrooms: Ringing, cheating, and classroom policies. *Communication Education, 55*(3), 280–294.

Campbell, S. (2008). Perceptions of mobile phone use in public: The roles of individualism, collectivism, and focus of the setting. *Communication Reports, 21*(2), 70–81.

Campbell, S. W., & Kwak, N. (2010). Mobile communication and civic life: Linking patterns of use to civic and political engagement. *Journal of Communication, 60*(3), 536–555.

Carey, J. (1983). Technology and ideology: The case of the telegraph. *Prospects, 8*, 303–325.

Caronia, L., & Caron, A. H. (2004). Constructing a specific culture: Young people's use of the mobile phone as a social performance. *Convergence: The International Journal of Research into New Media Technologies, 10*(2), 28–61.

CitySense. (2008). CitySense TM. Retrieved December 27, 2010 from http://www.city sense.com/.

De Gournay, C. (2002). Pretense of intimacy in France. In J. Katz, & M. Aakhus (Eds.), *Perpetual contact: Mobile communication, private talk, public performance* (pp. 193–205). Cambridge: Cambridge University Press.

de Souza e Silva, A. (2006). Interfaces of hybrid spaces. In A. Kavoori, & N. Arceneaux (Eds.), *The cell phone reader: Essays in social transformation* (pp. 19–43). New York: Peter Lang.

de Souza e Silva, A. (2008). Alien revolt: A case-study of the first location-based mobile game in Brazil. *IEEE Technology and Society Magazine, 27*(1), 18–28.

de Souza e Silva, A. (2009). Hybrid reality and location-based gaming: Redefining mobility and game spaces in urban environments. *Simulation & Gaming, 40*(3), 404–424.

de Souza e Silva, A., & Delacruz, G. (2006). Hybrid reality games reframed: Potential uses in educational contexts. *Games and Culture, 1*(3), 231.

de Souza e Silva, A., & Frith, J. (2010). Locative social mobile networks: Mapping communication and location in urban spaces. *Mobilities, 5*(4), 485–505.

de Souza e Silva, A., & Hjorth, L. (2009). Playful urban spaces: A historical approach to mobile games. *Simulation and Gaming, 40*(5), 602–625.

de Souza e Silva, A., & Sutko, D. M. (2008). Playing life and living play: How hybrid reality games reframe space, play, and the ordinary. *Critical Studies in Media Communication, 25*(5), 447–465.

de Souza e Silva, A., & Sutko, D. M. (2011). Theorizing locative technologies through philosophies of the virtual. *Communication Theory, 21*(1), 23–42.

de Souza e Silva, A., & Sutko, D. M. (Eds.). (2009). *Digital cityscapes: Merging digital and urban playspaces*. New York: Peter Lang.

Dede, C., Nelson, B., Ketelhut, D. J., Clarke, J., & Bowman, C. (2004). Design-based research strategies for studying situated learning in a multi-user virtual environment. Paper presented at the Proceedings of the 6th International Conference on Learning Sciences.

Delacruz, G. C., Chung, G. K. W. K., & Baker, E. L. (2009). Finding a place: Developments of location-based mobile gaming in learning and assessment environments. In A. de Souza e Silva, & D. Sutko (Eds.), *Digital cityscapes: Merging digital and urban playspaces* (pp. 251–268). New York: Peter Lang.

DoCoMo, N. (2002). NTT DoCoMo releases "open i-area" guidelines. From http://www.nttdocomo.com/pr/2002/000756.html.

DoCoMo, N. (2003). NTT DoCoMo to introduce first GPS handset. From http://www.nttdocomo.com/pr/2003/000943.html.

Dourish, P., Anderson, K., & Nafus, D. (2007). Cultural mobilities: Diversity and agency in urban computing. Paper presented at the Proceedings of the IFIP Conference Human-Computer Interaction.

Durham Peters, J. (2008). *Media and communications*. Boston: Blackwell Publishing Ltd.

Farley, T. (2005). Mobile phone history. *Telektronikk, 4*(3), 22–34.

Feiner, S., MacIntyre, B., Hollerer, T., & Webster, A. (1997). A touring machine: Prototyping 3D mobile augmented reality systems for exploring the urban environment. *Personal and Ubiquitous Computing, 1*(4), 208–217.

Fisher, C. S. (1994). *America calling: A social history of the telephone to 1940*. Los Angeles: University of California Press.

Flintham, M., Benford, S., Anastasi, R., Hemmings, T., Crabtree, A., Greenhalgh, C., et al. (2003). Where on-line meets on the streets: Experiences with mobile mixed reality games. Paper presented at the Proceedings of the SIGCHI Conference on Human Factors in Computing Systems.

Fortunati, L. (2002). Italy: Stereotypes, true and false. In J. Katz, & M. Aahkus (Eds.), *Perpetual contact: Mobile communication, private talk, public performance* (pp. 42–62). Cambridge: Cambridge University Press.

Foursquare. (2011). Wow! The foursquare community has over 10,000,000 members! From http://blog.foursquare.com/2011/06/20/holysmokes10millionpeople/.

Galloway, A. (2006). Locative media as socialising and spatialising practices: Learning from archaeology. *Leonardo Electronic Almanac, 14*(3/4), 12.

Gee, J. P. (2007). *What video games have to teach us about learning and literacy.* New York: Palgrave.

Gergen, K. (2002). The challenge of absent presence. In J. Katz, & M. Aakhus (Eds.), *Perpetual contact: Mobile communication, private talk, public performance* (pp. 227–241). Cambridge: Cambridge University Press.

Gergen, K. (2008). Mobile communication and the transformation of the democratic process. In J. E. Katz (Ed.), *Handbook of mobile communication studies* (pp. 297–310). Cambridge, MA: MIT Press.

Gergen, K. (2010). Mobile communication and the new insularity. *Interdisciplinary Journal of Technology, Culture and Education, 5*(1).

Geser, H. (2004). Towards a sociological theory of the mobile phone. *Swiss Online Publications.* Retrieved from http://socio.ch/mobile/t_geser1.htm.

Goggin, G. (2006). *Cell phone culture: Mobile technology in everyday life.* London, New York: Routledge.

Gordon, E. (2009). Redefining the local: The distinction between located information and local knowledge in location-based games. In A. de Souza e Silva, & D. M. Sutko (Eds.), *Digital cityscapes: Merging digital and urban playspaces* (pp. 21–36). New York: Peter Lang.

Groening, S. (2010). From 'A Box in the theater of the world' to 'The world as your living room': Cellular phones, television, and mobile privatization. *New Media & Society, 12*(8), 1331–1347.

Habuchi, I. (2005). Accelerating reflexivity. In M. Ito, D. Okabe, & M. Matsuda (Eds.), *Personal, portable, pedestrian: Mobile phones in Japanese life.* Cambridge, MA: MIT Press.

Hampton, K., & Gupta, N. (2008). Community and social interaction in the wireless city: Wi-fi use in public and semi-public spaces. *New Media and Society, 10*(8), 831.

Hampton, K., Livio, O., & Sessions, L. (2010). The social life of wireless urban spaces: Internet use, social networks, and the public realm. *Journal of Communication, 60,* 701–722.

Hemment, D. (2004). Locative dystopia 2. From http://www.makeworlds.org/node/76.

Hemment, D. (2006). Locative arts. *Leonardo, 39*(4), 348.

Hewlett-Packard, D. L. (2005). Mscape.

Hight, J. (2006). Views from above: Locative narrative and the landscape. *Leonardo Electronic Almanac, 14*(7/8), 9.

Höflich, J. (2006). The mobile phone and the dynamic between private and public communication: Results of an international exploratory study. *Knowledge, Technology & Policy, 19*(2), 58–68.

Humphreys, L. (2005). Cellphones in public: Social interactions in a wireless era. *New Media & Society, 7*(6), 810–833.

Jacobs, J. (1961). *The death of life of great American cities.* New York: Random House.

Kafai, Y. B. (2006). Playing and making games for learning: Instructionist and constructionist perspectives for game studies. *Games and Culture, 1*(1), 36.

Katz, J. (2005). Mobile phones in educational settings. In K. Nyiri (Ed.), *A sense of place: The global and the local in mobile communication* (pp. 305–317). Vienna: Passagen.

Katz, J., & Aakhus, M. (Eds.). (2002). *Perpetual contact: Mobile communication, private talk, public performance*. Cambridge: Cambridge University Press.

Kellerman, A. (2006). *Personal mobilities*. London: Routledge.

Knowlton, J., Spellman, N., & Hight, J. (Artist). (2002). *34 North 118 West*.

Laurier, E. (2001). Why people say where they are during mobile phone calls. *Environment and Planning D, 19*, 485.

Licoppe, C., & Guillot, R. (2006). ICTs and the engineering of encounters. A case study of the development of a mobile game based on the geolocation of terminals. In J. Urry, & M. Sheller (Eds.), *Mobile technologies of the city* (pp. 152–163). New York: Routledge.

Licoppe, C., & Heurtin, J.-P. (2001). Managing one's availability to telephone communication through mobile phones: A French case study of the development dynamics of mobile phone use. *Personal and Ubiquitous Computing, 5*(2), 99–108.

Licoppe, C., & Heurtin, J.-P. (2002). France: Preserving the image. In J. E. Katz, & M. Aahkus (Eds.), *Perpetual contact: Mobile communication, private talk, public performance* (pp. 94–109). Cambridge: Cambridge University Press.

Licoppe, C., & Inada, Y. (2006). Emergent uses of a multiplayer location-aware mobile game: The interactional consequences of mediated encounters. *Mobilities, 1*(1), 39–61.

Licoppe, C., & Inada, Y. (2009). Mediated co-proximity and its dangers in a location-aware community: A case of stalking. In A. de Souza e Silva, & D. M. Sutko (Eds.), *Digital cityscapes: Merging digital and urban playspaces* (pp. 100–128). New York: Peter Lang.

Ling, R. (2004). *The mobile connection: The cell phone's impact on society*. San Francisco: Morgan Kaufman.

Ling, R. (2008). *New tech, new ties: How mobile communication is reshaping social cohesion*. Cambridge, MA: MIT Press.

Ling, R., & Yttri, B. (1999). Nobody sits at home and waits for the telephone to ring: Micro and hyper-coordination through the use of the mobile telephone. Paper presented at the Perpetual Reachability Conference.

Ling, R., & Yttri, B. (2002). Hyper-coordination via mobile phones in Norway. In J. Katz, & M. Aakhus (Eds.), *Perpetual contact: Mobile communication, private talk, public performance* (pp. 139–169). Cambridge: Cambridge University Press.

Mante, E. (2002). The Netherlands and the USA compared. In J. E. Katz, & M. Aahkus (Eds.), *Perpetual contact: Mobile communication, private talk, public perfomance* (pp. 110–125). Cambridge: Cambridge University Press.

Matsuda, M. (2005). Mobile communication and selective sociality. In M. Ito, D. Okabe, & M. Matsuda (Eds.), *Personal, portable, pedestrian: Mobile phones in Japanese life* (p. 123). Cambridge, MA: MIT Press.

Meyrowitz, J. (1985). *No sense of place: The impact of electronic media on social behavior*. New York: Oxford University Press.

Monk, A., Carroll, J., Parker, S., & Blythe, M. (2004). Why are mobile phones annoying? *Behaviour & Information Technology, 23*(1), 33–41.

Monk, A., Fellas, E., & Ley, E. (2004). Hearing only one side of normal and mobile phone conversations. *Behaviour & Information Technology, 23*(5), 301–305.

Moores, S. (2004). The doubling of place: Electronic media, time-space arrangements and social relationships. In B. Couldry, & A. McCarthy (Eds.), *Media/Space: Place, scale and culture in a media age* (p. 21). London: Routledge Comedia Series.

Mooser, J., Wang, L., You, S., & Neumann, U. (2007). An augmented reality interface for mobile information retrieval. Paper presented at the third IEEE International Conference on Multimedia and Expo.

Nickerson, R. C., Isaac, H., & Mak, B. (2008). A multi-national study of attitudes about mobile phone use in social settings. *International Journal of Mobile Communications*, *6*(5), 541–563.

Okabe, D., & Ito, M. (2005). Keitai and public transportation. In M. Ito, D. Okabe, & M. Matsuda (Eds.), *Personal, portable, pedestrian: Mobile phones in Japanese life* (pp. 205). Cambridge, MA: MIT Press.

Paulos, E., & Goodman, E. (2003). *The familiar stranger: Anxiety, comfort, and play in public spaces*. Berkeley, CA: Intel Corp.

Plant, S. (2001). *On the mobile: The effects of the mobile phone on social and individual life*. London: Motorola, Inc.

Polak, E., & the Waag Society (Artist). (2002). *Amsterdam Real Time*.

Proboscis (Artist). (2002–2004). *Urban Tapestries* (Mobile Annotation).

Puro, J. P. (2002). Finland, a mobile culture. In J. Katz, & M. Aakhus (Eds.), *Perpetual contact: Mobile communication, private talk, public performance* (pp. 19–29). Cambridge: Cambridge University Press.

Rheingold, H. (2002). *Smart mobs: The next social revolution*. Cambridge, MA: Perseus Publishing.

Rogers, E. (1995). *Diffusion of innovations*. New York: Free Press.

Rueb, T. (Artist). (2005). *Itinerant* (Sound Installation).

Schivelbusch, W. (1986). *The railway journey: The industrialization of time and space in the 19th century*. Berkeley, CA: University of California Press.

Sheller, M., & Urry, J. (2003). Mobile transformations of 'public' and 'private' life. *Theory, Culture, and Society*, *20*(3), 107–125.

Sickhur, K., & Smith, A. (2010). 4% of online Americans use location-based services. From http://www.pewinternet.org/Reports/2010/Location-based-services.aspx.

Sotamaa, O. (2002). All the world's a Botfighters stage: Notes on location-based multiuser gaming. Paper presented at the Proceedings of Computer Games and Digital Cultures Conference.

Sparacino, F. (2002). The museum wearable: Real-time sensor-driven understanding of visitors' interests for personalized visually-augmented museum experiences. Proceedings of Museums and the Web (WM 2002) Boston.

Sutko, D. M., & de Souza e Silva, A. (2011). Location aware mobile media and urban sociability. *New Media & Society*, *13*(5), 807–823.

Swartz, J. (2011). Tech titans crash start-ups' party at SXSWi fest. *USA Today*, March 12. Retrieved from http://www.usatoday.com/money/media/2011–03–11-Sxswnewbies 11_CV_N.htm.

Tuters, M., & Varnelis, K. (2006). Beyond locative media: Giving shape to the internet of things. *Leonardo*, *39*(4), 357–363.

Umble, D. Z. (2003). Sinful network or divine service: Competing meanings of the telephone in Amish Country. In L. Gitelman, & G. Pingree, B. (Eds.), *New media: 1740–1915* (pp. 139–156). Cambridge, MA: MIT Press.

Urry, J. (2000). *Sociology beyond societies: Mobilities for the twenty-first century*. New York: Routledge.

Vygotsky, L. S. C. M. (1978). *Mind in society: The development of higher psychological processes*. Cambridge, MA: Harvard University Press.

Weilenmann, A. (2003). I can't talk now, I'm in a fitting room: Availability and location in mobile phone conversations. *Environment and Planning A*, *35*(9), 1589–1605.

Wellman, B. (2002). Little boxes, glocalization, and networked individualism. Paper presented at the Revised Papers from the Second Kyoto Workshop on Digital Cities II, Computational and Sociological Approaches.

Whyte, W. (1980). *The social life of small urban spaces* Washington, DC: Conservation Foundation.

Wiley, M. M., & Rice, S. (1933). *Communication agencies and social life*. New York: McGraw-Hill.

Wilson, J. (2006). 3G to Web 2.0? Can mobile telephony become an architecture of participation? *Journal of Research into New Media Convergence: The International Technologies*, *12*(2), 229–242.

Wood, J. (Artist). (2000). *GPS Drawings*.

Xiong, L., Ratan, R., & Williams, D. (2009). Location-based mobile games: A theoretical framework for research. In A. de Souza e Silva, & D. M. Sutko (Eds.), *Digital cityscapes: Merging digital and urban playspaces* (pp. 37–54). New York: Peter Lang.

SECTION II

Location-awareness in the Contemporary City

4
LOCATIONAL PRIVACY

As she approaches a coffee shop, Johanna's mobile phone vibrates in her pocket. She thinks: "Why is somebody texting me here? Don't they know I'm out of the country?" She is surprised when she sees that the text message is from someone called "Rick's Tap and Grill," which is the bar right across the street from the coffee shop. Out of curiosity, she clicks on the message. It tells her that if she comes in, she'll get a free drink. It's too early for a drink so she isn't tempted by the offer, but she does think to herself, "How do they possibly know that I'm here?"

She enters the coffee shop, gets a table and a cup of coffee. After a few minutes, she looks at her mobile phone, opens *Foursquare* and checks where her friends had checked in over the past 48 hours since she started traveling. "It is interesting to use *Foursquare* from a distance," she thinks, almost relieved that she doesn't feel the pressure to acknowledge friends just because they are close by. She also has just downloaded *LooptMix* and decides to look at the application out of curiosity, to see if there is anyone checked in nearby. To her surprise, there were actually five users in a very close range. Before she starts clicking on one of their profiles, however, she gets a message from another user that asks: "Would you like to talk?" She feels very awkward by that impromptu approach and for a moment she doesn't know how to react. She looks around, suspicious, and realizes she is not in the mood to meet anybody new. Afraid that somebody might approach her, she gets up and leaves the coffee shop.

In 2006, Kevin Kelly estimated that the whole of humanity's existing published knowledge could be digitized onto a 50 petabyte disk. A single petabyte is an incredible amount of data, but far more remarkable than Kelly's 2006 estimate is the rate of change since 2006. Two years later, a group of Google engineers published a white paper that showed that Google processed 20 petabytes of data *a day* (Dean & Ghemawat, 2008). Google is not the only company dealing with massive amounts of data: AT&T carries 16 petabytes of data and IP traffic a day, "the equivalent of a 2.5-megabyte music download for every man, woman and child on the planet" (AT&T, 2008).

The sheer amount of data produced daily is astounding, certainly one of the more amazing byproducts of what Castells (2000) calls the *information age*. But when numbers such as those cited above are thrown around in conversation, people too often forget the contents of much of this data: personal information. Individuals fill out online forms to get drivers' licenses, loyalty shopping cards, memberships to discussion forums, and names on conference schedules. All these forms produce more data. Many daily activities are recorded as well. Web history can be matched to IP addresses to produce personalized data; credit card companies record customers' transactions; websites such as Amazon compile records of reading history. Even grocery stores, through their loyalty shopping programs, catalog many individuals' purchases. The increasing use of mobile phones, especially text messaging, has also resulted in "the explosion of mobile data" (Hashimoto & Campbell, 2008, p. 538). And the amount of networked data is only going to increase in the future with the move to cloud computing.

Cloud computing is the idea that access to information is moving from the desktop to the "webtop," that is, people will increasingly store and interact with data that is networked and accessed via the web, rather than on local computer hard drives (Andrejevic, 2007). Cloud computing is not the only technological development that will increase the amount of recorded personal data. Mobile technologies will as well, especially through location-aware capabilities. Many of today's mobile phones feature location-awareness, and as the iPhone App store shows, there has been an increasing number of location-based applications that take advantage of these capabilities.

As we have seen in the previous chapter, location-based applications have only begun to gain popularity since 2008 when they started being used more frequently by the general public. The lack of popularity in the past can be partly explained by the small number of location-aware devices capable of running useful LBS, and by smaller developers' lack of opportunity to develop their own location-based applications. With the creation of the iPhone App store in 2008, that changed, and there are now myriad mobile devices that are capable of running LBS. However, there is still some hesitance to develop and adopt LBS, which is most related to concerns about locational privacy. As LBS gain popularity, these privacy concerns increase.

Often media outlets and software designers address privacy as if it is an unambiguous term, but privacy is in fact a very complex issue. Daniel Solove (2008) shows that the majority of definitions of privacy are either too broad or too narrow and that privacy issues need to be understood contextually. Rather than trying to propose yet another general definition of privacy, in this chapter we conceptualize privacy in the context of location-aware mobile interfaces. By examining privacy concerns contextually, we are better able to understand what is driving these concerns and discuss their applicability to location-aware devices. Acknowledging that conceptions of privacy are both socially constructed and influenced by technological development, this chapter investigates the privacy issues raised by the adoption of location-aware interfaces.

We begin this chapter by discussing electronic databases and online services, examining how their emergence has raised new privacy issues. Then we move on to develop a conceptual framework for understanding privacy in the context of location-aware mobile interfaces. We highlight how locational privacy is different from privacy issues already addressed by critics of the World Wide Web. When location-aware devices become the interface to access both the Internet and public spaces, existing issues of digital privacy acquire an additional element: location. Location information raises important issues of surveillance, and we show that the fears of surveillance most commonly associated with the use of LBS break down along three main categories: governmental surveillance, corporate surveillance, and interpersonal surveillance. We analyze the first two as forms of "top-down" surveillance, while the last one represents what we call "collateral" surveillance.[1]

A central point to our argument is the notion that privacy is not some "thing" out there waiting to be violated. As with the notions of public and private, privacy is socially constructed, and it shifts across historical periods and across cultures. Privacy is also closely related to technological development. Because it is a social construct, and because the social cannot be separated from the technological (Castells, 2000; Latour, 1987, 2005), the technologies we use every day affect the meaning of privacy. We thus conclude this chapter by discussing how the adoption of location-aware mobile technologies may signal a shift in our understanding of privacy. In the context of location-aware technologies, issues of privacy are intimately related to ideas of control. The meaning of locational privacy is directly influenced by the ability to control location information.

Online Privacy and Electronic Databases

Expectations of privacy have changed through the years, but that does not mean that privacy is not worth protecting, or at least that certain understandings of privacy are not worth protecting. The fight to protect privacy became increasingly important with the growth of electronic databases. The two decades after the invention of the World Wide Web have been filled with an exponential increase in the amount of information recorded about us. As Solove (2001) writes, "The small details that were once captured in dim memories or fading scraps of paper are now preserved forever in the digital minds of computers, vast databases with fertile fields of personal data" (p. 1394). Because personal information has become such a valuable commodity, the growth of databases has led to major privacy concerns discussed in the popular press (Cauley, 2006), post-structural theory (Poster, 1990), legal theory (Solove, 2004, 2008), surveillance studies (Haggerty & Ericson, 2000; Lyon, 1994, 2006b), and many other fields.

Databases thrive on information, and many of the most popular web services are built on the trade of information. We are offered rewards for signing up for services and providing personal information such as our email address, our physical address, our name, and our telephone number. And because information is traded among databases, our data exist simultaneously in multiple places. We contribute to the increasing amount of data about ourselves stored in electronic databases nearly every day as more and more of our actions are recorded by websites, surveillance cameras, or print forms we sign. One of the major problems with electronic databases, according to Solove (2004), is that most of us have little understanding about what information is out there in databases, who has access to that information, or if there is any recourse against the collection of the information. Furthermore, although people might be aware of some ways of protecting their personal information online (e.g., by not using an open wireless network), most people are unaware of hidden mechanisms used by major companies to retrieve personal information. Intel Labs researchers Sunny Consolvo et al. (2010) developed a software named the *Wi-Fi Privacy Ticker* that runs in the background of regular browsers and through a graphic interface "displays information about the exposure of sensitive terms that are sent to and from a user's computer" (p. 1). The software also prevents the unencrypted transmission of sensitive terms (such as passwords and usernames) from users' computers. The goal of the *Wi-Fi Privacy Ticker* is to expose how websites often share users' personal information with third parties without users' previous consent or awareness. Because sites often share information without users' knowledge, most people have little means to protect what is done with their personal data.

Not all losses of privacy, however, can be attributed to a lack of knowledge. David Nguyen, Alfred Kobsa and Gillian Hayes (2008) found that while people do express concern about their lack of control over their personal information in a general sense, they do not typically care that their credit card company records all their purchases or that Amazon stores their reading history. So, while people may bemoan the loss of privacy, they often cooperate with the practices that allow personal information to circulate in databases, in part because many people believe the benefits outweigh the possible consequences.

However, as we will see later, many users are also unaware of the consequences of sharing personal information online. An example of the possible consequences of the mass collection of personal information in databases is the AOL search history fiasco. In 2006, AOL released the anonymous search histories of 656,000 AOL users so that people could use the data for research purposes (Halavais, 2009). The data included no personal information, but because the search queries were grouped by user, *New York Times* writers Mark Barbaro and Tom Zeller Jr. (2006) were able to identify a specific user. They did so by analyzing some of her specific search terms, which made it fairly easy to identify who was doing the searching. Each individual search string had little meaning and would have remained anonymous forever. When the different strings were combined, however, they painted a robust picture of the woman. AOL was immediately criticized for releasing the data and pulled it down two months later.

A substantial portion of the debate about privacy and information collection online has centered on social networking sites (SNS), such as *Facebook*, *MySpace*, *Orkut*, and *Friendster* (Hargittai and boyd, 2010; boyd, 2008, 2010). *Facebook* in particular has often been targeted by the popular press and privacy advocates. *Facebook* is notorious for changing its user agreement and making previously obscure information public. Users who consent to one set of *Facebook* privacy policies then watch as the policies are changed without their consent. Information is valuable, and the more information people share on *Facebook*, the more value the service has. For that reason, it makes sense economically for *Facebook* to change its privacy policy to encourage people to share more information. Sometimes these changes go more or less unnoticed, and sometimes they cause outrage from users and critics.

One of the most famous examples of a privacy uproar over *Facebook* came in 2006 when the website introduced the News Feed feature. The News Feed has become the default "face" of *Facebook*. When users log in, the first thing they see is a stream of their friends' recent activity. For example, they see that one friend recently changed his relationship status to single and that another friend recently posted condolences on his Wall. For the more than 400 million users who have joined *Facebook* since the creation of the News Feed, this is simply the *Facebook* they know. But for many of the people who were on *Facebook* when the News Feed was introduced, this was a major privacy controversy.

The News Feed did not make private information public in the simple sense. Everything published in a News Feed was already publicly available on the user's *Facebook* profile. People had already consented to make this information available to their friends. For example, while the News Feed might publish that a friend recently changed his political party to Democrat, before the News Feed was introduced his friends could still see that he changed his political status; they just would have had to go to his profile to see his listed political party. So with the News Feed nothing that was private became public, and yet users still felt their privacy was invaded by *Facebook*'s change.

The problem with *Facebook*'s introduction of the News Feed was that privacy perceptions are contextual. One of the challenges that users of social networking sites face is that these sites present a new context for information sharing that people had not dealt with before. Most people know that certain behaviors are acceptable in private and not acceptable in public spaces, but with social networking sites there was (and still is) confusion about what types of spaces these sites are supposed to be. With the News Feed, the context in which users shared information changed and began being broadcast to friends through updates. As a result, information that was never really private was then broadcast in new ways without users' consent.

As danah boyd (2008) argues, the News Feed example shows that we cannot view information as a binary code of 0 = private and 1 = public. Take the discussion of databases above. Many people would probably not care about disclosing their zip code, or the name of the elementary school they attended, or even the chain of grocery stores where they shop. But when these discrete pieces of information are collected and combined with each other, a privacy problem may arise. Relational databases take these discrete pieces of information and aggregate them in order to paint what is often a fairly robust picture of the individual, a picture Solove (2004) calls "the digital person." These digital identities are used to predict shopping behavior, patterns of mobility, or trigger criminal records. Privacy is violated because information that was shared in one context is combined with other information in a way that is removed from the original context and can potentially harm the individual. In a similar way, *Facebook*'s News Feed takes pieces of information that were previously separate and puts them all together. It changes the context of the information (boyd, 2008).

Facebook is certainly not the only example of this. boyd also points to an earlier example when in 1995 DejaNews developed a tool that let people search through UseNet postings. UseNet postings were always public in a sense, but they were not searchable and were often very specialized, so the typical UseNet user was not likely to stumble upon a random posting. When DejaNews made the boards searchable, that all changed. The postings then showed up in searches and were often removed from their original context. While the original posting were never "private," they were not searchable, so the only people who would find a post were people interested enough to be scrolling through posts on that topic. But when posts began showing up in decontextualized searches, the user community felt its privacy was violated because they had not given their consent to show up in search results and felt they had lost some of the anonymity and control enabled by the previous information architecture of UseNet. DejaNews, like *Facebook* more than ten years later, failed to see that information is not either private or public, it is often both and neither (boyd, 2008).

Paradoxically, it is also possible to look at the *Facebook* News Feed example from another angle: as an example of how quickly understandings of privacy can shift with technological (or in this case application) development. The idea of following everyone else's actions was quickly accepted, and most users did not stop using *Facebook* over the privacy flap. More recently, in December 2009, *Facebook* made some major changes to users' privacy settings, by making users' personal information publicly available by default to everyone on the Internet. The policy changes led to an uproar in the popular press and in certain communities. A movement called "Quitfacebookday" encouraged people to band together to quit *Facebook* on May 31, 2010. The movement gained attention from prominent sources such as *Wired* and *TechCrunch*, but ultimately failed to make much of an impact. No one can be sure how many people quit, but most estimates put it at a lower number than the number of people who join *Facebook* daily.

We are not arguing that large entities such as *Facebook*—or for that matter any technology or corporation—have full power to dictate privacy shifts. But often the development of new technologies both reflects and influences existing privacy concerns, as we saw with the case of the Kodak camera in Chapter 2. *Facebook* has certainly lost battles in the privacy wars. *Facebook*'s 2007 rollout of Beacon— a partnership between *Facebook* and 44 partner websites and a key part of its advertising plan—was a disaster. When people did things on partner sites such as purchase products, these purchases were published to their friends' News Feeds. The most famous example of how Beacon violated users' privacy came from a man named Sean Lane who had bought his girlfriend an engagement ring on Overstock.com and only noticed later that the purchase had been broadcast to all his friends (Nakashima, 2007). Users revolted, and what was supposed to be a key plank in *Facebook*'s effort to monetize was shut down less than two years later (Perez, 2009).

Google Buzz was another example of people reacting negatively to what was perceived as an invasion of users' privacy. *Google Buzz* took users' *Gmail* contacts and made them members of their *Buzz* social network by default. The designers did not realize that people who email each other frequently do not necessarily want to be in the same social network, or to read each other's status updates about personal affairs. *Google* was sharply criticized for failing to understand that email quantity does not equal close friendship. The initial negative reaction was likely a contributing factor in keeping *Buzz* from growing into a hugely popular service.

What we can see from the examples above is that privacy issues are often double-sided. On the one hand, people have pre-conceived notions of privacy. If services are perceived to violate privacy, people often fight back by protesting or not adopting these services. On the other hand, frequently technologies that seem to disrupt the social fabric then become part of the social fabric, therefore also influencing existing notions of privacy. Something similar is happening with the development of location-aware technologies and emerging issues of locational privacy.

Locational Privacy

Most of the examples discussed above were taken from the perspective of users who connect to the Internet via a desktop or laptop computer. In these cases, the users' locations did not influence the type of data they access. For example, someone's *Facebook* page will look similar whether the user is in the United States or Denmark. And although most web searches from desktop computers retrieve users' IP addresses, when location-aware mobile technologies start to be used as interfaces to connect to the Internet, the user's location becomes a crucial determinant of the type of data accessed. Consequently, privacy issues become more directly interconnected with location.

Sharing information online can certainly affect one's offline interactions. An unfortunately tagged photo on *Facebook* can lead to a loud fight in a dorm room, or to the tagged person being fired from a job (Stross, 2007). But, as we explained in Chapter 2, contemporary accounts of public spaces should take into consideration the idea of *net locality* (Gordon & de Souza e Silva, 2011), according to which digital information becomes an intrinsic part of people's experience of urban spaces. Within net locality, it is no longer possible to address distinctions between physical and digital spaces, and users' locations become an important part of their networked interactions, influencing the people they meet and the information they access. For example, when someone standing on a street uses a location-based service (LBS) to access information about a specific location, the information looked up through the mobile phone interface becomes part of that location in the same way a street sign, a flyer, or graffiti is part of the location. Often, the information people share, either with the service or with other users, is tied back to the location of their corporeal body. It is important to remember, however, that to use a location-based service people have to disclose their location information to mobile phone providers, and the disclosure of personal location information often raises privacy concerns. The privacy concerns that emerge with disclosing location information have been called "locational privacy" (Blumberg & Eckersley, 2009).

Issues raised by locational privacy are nothing new, as Mark Monmonier (2002) shows in his book *Spying with Maps*. Monmonier discusses how certain practices we typically accept as commonplace reconfigured ideas of locational privacy. One example was the beginning of aerial mapping practices by the United States Agriculture Department during the 1930s. The Agriculture Department flew planes over people's land to ensure that they were in compliance with land and conservation laws, basically enacting a huge surveillance structure that allowed for the monitoring of fields. Monmonier notes that what was remarkable about the growth of aerial mapping was that, despite its potentially intrusive nature, few popular concerns were raised. Monmonier (2002) goes on to argue that locational privacy has only recently arisen as a major concern. Until "one's whereabouts was so easily determined, archived, and sold, locational nakedness was hardly an issue" (p. 175).

Locational privacy concerns can be traced back to two major developments: the growth of databases as discussed above, and the popularization of technologies that can track users' location. The most prominent locative technology has been GPS, at least since the U.S. government stopped degrading the satellite signal in 2000. However, individuals' mobile phones could be located before phones began including GPS. As we mentioned in the previous chapter, governments, through triangulation of waves, could determine someone's position based on the location of a mobile phone in relation to the nearest cell towers, which was central to the Enhanced 911 (E911) system.

The E911 system was widely accepted as necessary, but privacy advocates immediately raised concerns about who could access mobile phones' location. The fear was that the location information would not only be available to emergency units but also to law enforcement agencies. *Wired* writer Chris Oakes (1998) wrote an article warning users that their mobile phones were being turned into tracking devices, and advocates such as James Dempsey, a representative of the Center for Democracy and Technology, urged the FCC to clarify when law enforcement agencies could access a mobile phone's location. Of course, the debate about E911 occurred before the 9/11 attacks of 2001, and discussions about the powers of law enforcement to invade privacy have changed since then. The E911 rollout did presage the privacy issues and concerns surrounding new GPS-enabled mobile phones turning into "tracking systems." A major difference, however, from early location tracking systems such as E911 and the current LBS is that with E911 location was disclosed to emergency services to increase safety and security. With most commercial LBS, location is traded for convenience, entertainment, and profit.

The fact that so many people are willing to share their location and be surveilled for the convenience of using LBS may, in itself, reflect a cultural shift. In his work on social control and surveillance, William G. Staples (1997) discusses how surveillance has become ubiquitous and also normalized and accepted. Reality television turns surveillance into entertainment, stores and streets include countless security cameras, and we are offered special deals all the time in exchange for our personal information. Staples uses many small examples of widely accepted surveillance to argue that the public is increasingly accepting being watched most of the time. According to his framework, people are willing to let their location be tracked in exchange for convenience and entertainment.[2] Without the normalization of all these types of micro-surveillance, people would not be willing to let companies and friends have access to their location information. However, just because people are more accepting of being watched, it does not mean that they do not care about privacy.

As with issues of online privacy, locational privacy needs to be understood contextually. Location is not inherently private. We do not have our privacy violated every time we share our location with a service or a friend, just like it is not a privacy violation if someone tells her friend she is at a cafe. The loss of privacy occurs when the context shifts away from how the information was originally intended. There is a general expectation that the public dissemination of our location will only occur with our consent and that others are not tracking our location history for their own benefit. When these expectations are violated, concerns about privacy emerge.

Expectations of privacy are ultimately about control, often control over one's ability to maintain a certain level of anonymity. As Daniel M. Sutko and Adriana de Souza e Silva (2011) suggest,

> The ability to remain anonymous as part of the mass [in a city] is one reason why anonymity is so central to urban sociology and one reason why LMSN[3] applications seem so threatening to privacy. The ability to identify individuals within the mass bypasses the *blasé* attitude theorized by Simmel as a response to the overwhelming stimuli of the city and the masses within.
>
> (p. 814)

As we have seen in previous chapters, the *blasé* attitude works as a social shield that helps individuals remain anonymous in the crowd. It is a means of feeling comfortable in public spaces by creating a distance from them and controlling other people's access to the personal self. When this anonymity is broken, people might feel threatened, and perceptions of an invasion of privacy begin to emerge. A notable example happens in the movie *Minority Report* when Tom Cruise's character enters *The Gap* and is immediately greeted by his name. The electronic salesperson suggests products based on his previous purchases and his personal information file in the store's database. While just walking into a store is not perceived as an invasion of privacy, the movie portrays a dystopic future in which privacy is non-existent and store computers know exactly when people enter, they identify people immediately, and then make predictions based on their previous purchases. As Adam Greenfield (2006) argues, the central concern about these information-rich ubiquitous computing scenarios is privacy.

However, understanding concerns about locational privacy is not as simple as arguing that each time a person is located their privacy is violated. While we expect to be anonymous in some situations, there are other cases in which we might want to be found. Active *Foursquare* users, for example, want to share their location with other members of their social network. In order to do that, they need to sign up to use the application and give consent for their location to be tracked by the mobile phone provider and by other players. At the same time though, players may think their privacy is invaded if they receive location-based advertisements for which they did not sign up.

To better understand the major contextual issues surrounding locational privacy triggered by the use of location-aware mobile interfaces, we performed a study on four months of media discourses on location-based services (de Souza e Silva & Frith, 2010a). Our data revealed that issues of locational privacy are generally associated with fears of surveillance: "top-down" forms of surveillance in the case of governmental and corporate surveillance, and "collateral" surveillance, in the case location is disclosed to other people.

Governmental Surveillance

The fear of government surveillance, while expressed fairly often in the media, is likely the least common privacy concern expressed by users of location-based services because, as research shows, many people believe they have nothing to hide from the government (Solove, 2007). However, concerns about governmental surveillance are still important and worth discussing. These fears expressed in the media about governmental surveillance refer to the increasingly prevalent collection of location data by mobile phone carriers. Carriers record users' GPS location, and their privacy policies often say little about when or if they are willing to turn that information over to the government. As we will discuss below, the major U.S. phone carriers will turn over personal information (including location information) to the government.

Most commonly, fears expressed about governmental surveillance and location-aware mobile technologies adopt the implicit Orwellian metaphor of Big Brother, a metaphor that has so strongly shaped understandings of privacy that Solove (2001, 2004) argues that even when people do not explicitly mention Big Brother or Orwell, they still draw from a privacy framework conceptualized through the metaphor of Orwell's famous dystopian novel *1984*. With location-aware mobile technologies, the link to Big Brother is obvious. This fear traces back to the earlier debates about E911, where privacy advocates feared that law enforcement agents and the government were turning mobile phones into tracking devices. With cellular triangulation, however, the best law enforcement agencies could do was locate individuals within 100 meters. But with new GPS-enabled mobile phones and other GPS tracking devices, people can now be located as close as five meters on a clear day.

The thought of a government being able to locate its citizens so precisely is worrisome to some people, and it was one of the main concerns identified in the Electronic Frontier Foundation's (EFF) policy paper "On locational privacy, and how to avoid losing it forever" (Blumberg & Eckersley, 2009). The EFF approached locational privacy issues from a standpoint that implicitly draws from the Big Brother metaphor. To frame the importance of locational privacy, the document asked a series of questions about information individuals would not want to be accessible, such as: "Did you go to an anti-war rally on Tuesday?" "Did you go to a small meeting to plan the rally the week before?" "Did you walk into an abortion clinic?" (Blumberg & Eckersley, 2009, n.p.)

These questions imply that users of location-aware technologies should fear an intrusive government that cares about who went to the anti-war rally and who has been to an abortion clinic. The EFF is not alone in linking the adoption of location-aware devices to an Orwellian future where the government can constantly monitor our location. Popular press sources as diverse as the *Guardian* and *E-week* have also published articles about how location-aware technologies contribute to our move toward an "Orwellian state" (de Souza e Silva & Frith, 2010a).

While Solove (2008) points out that the metaphor of Big Brother has shaped late twentieth-century legal thought, he argues that the Big Brother metaphor is not analytically useful. His arguments are echoed by other artists and scholars who focus on the social uses of location-aware interfaces. Matt Adams, member of the British artist group Blast Theory, for example, argues that "we are still locked in an Orwellian paradigm that has long since passed its sell by date" (de Souza e Silva & Sutko, 2009, p. 81). Adams' argument is that the surveillance structure people fear when they conceptualize the privacy issues of location-aware mobile interfaces inside the Big Brother metaphor no longer exists. Instead, the all-seeing surveillance structure imagined inside that framework has been replaced (if it ever existed) by capillary forms of surveillance. The major fear with location-aware mobile interfaces should instead be focused on both businesses and other users who can invade one's privacy by accessing location information. Other scholars have made similar points, arguing that the Big Brother metaphor misses too much (Lyon, 2006a). We agree that privacy in the context of location-aware interfaces does need to move past an understanding that relies too heavily on the idea of a coherent surveillance system run by governments; however, some of the concerns captured in the metaphor of Big Brother still apply to locational privacy.

In late 2009, Chris Soghoian released data showing that Sprint-Nextel had turned over customers' GPS information to law enforcement official eight million times between September 2008 and October 2009 (Zetter, 2009). We also should not forget that bills such as the Patriot Act grant the U.S. government a far greater license to access personal information in a variety of contexts, including, most notoriously, wiretapping without a warrant. In a national landscape shaped deeply by 9/11 and the interminable War on Terror, governments are increasingly interested in personal data, and location-aware mobile interfaces provide one of the most valuable types of data: location. It is important to note, however, that fears of governmental surveillance are also not homogenous across the world. For example, while U.S. citizens typically have low confidence in the government (Lock, Shapiro, & Jacobs, 1999), other countries such as Sweden and Denmark report a high level of trust in their governments (Berggren, Elinder, & Jordahl, 2008; Rothstein & Uslaner, 2005). These different levels of trust likely influence how users feel about governmental agencies' ability to access location information. However, even when sharing location information with the government is not viewed as an immediate issue, corporations are also the target of privacy concerns, especially when it comes to location-based advertising.

Corporate Surveillance and Location-based Advertising

Some of the largest Internet companies have made their fortunes off personal information. Google, for example, is ultimately an advertising company. Google revolutionized Internet search, released a superior email client users flocked to, and designed the open source mobile OS Android that is currently competing with BlackBerry and the iPhone, but Google makes most of its money off its ability to provide people with ads that are contextually relevant. To provide contextually relevant advertisements requires a certain amount of personal information. When someone sends an email through *Gmail*, the text of the email is automatically parsed, and advertisements that are perceived as relevant appear on the *Gmail* screen. This is an example of how relevant personal information can be for advertisers. With location-awareness, advertisements now can be targeted based on people's location.

Location-based advertising (LBA) is a fairly simple concept. People sign up for a service that provides coupons to a certain store, and when they walk past the store, the service lets them know if there are any special offers. The value of LBA is obvious. People can receive offers at the moment they are most likely to use them: when passing by the store. Rather than clipping out coupons at home and maybe remembering to use them later, people instead receive coupons based on their location. LBA is a form of surveillance, for better or worse. For a company to target people with a locative ad, it has to know where they are. Kolmel and Alexakis (2002) identify two approaches to LBA: push and pull. Push refers to services that actually send offers to users' phones. A pull method is seen when, for example, a *Foursquare* user logs into the application, looks up places nearby, and sees that the Mexican restaurant down the street is offering a free appetizer. Within the push method, there are two approaches to delivering advertisements: not requested or indirectly requested. The not requested approach sends users ads in a manner similar to spam email. The indirectly requested requires users' agreement to receive advertisements. Just like with web services, however, users do not have to agree to receive advertisements directly from the company. Users can sign up for one service that then will provide information to another company.

One of the problems with any kind of advertising targeting mobile phones is that most people sees their mobile phones as a more personal device than, say, a desktop computer (Fortunati & Cianchi, 2006; Ito, Okabe, & Matsuda, 2005). Receiving an unrequested ad on a mobile phone is then perceived as far more intrusive than a *Facebook* ad on the periphery of one's Internet browser. Some of the more perceptive mobile marketing scholarship recognized this problem (Barnes and Scornavacca, 2004; Leppaniemi and Karjaluoto, 2005), and warned marketers that people's attachment to their mobile phone may cause them to feel their personal space had been violated by a poorly targeted, irrelevant ad.

With LBA, every time an individual receives an ad, they are reminded that a company has access to their location. These reminders have contributed to a fear about corporate surveillance that has been discussed fairly frequently in the popular press. For example, the *Guardian* featured an article that claimed that "People are being told that they are signing up for marketing when in fact they are being opted into a massive surveillance strategy" (Warren, 2009). Many of our actions can be tracked on the Web, such as our browsing history, time spent on certain websites, and book searches. With location-aware mobile interfaces, companies can now extend their reach outside of the computer screen. They can track individuals' location if they are using LBS that are affiliated with that company.

The privacy and surveillance implications of LBA need to be discussed further. LBA is no doubt a surveillance strategy, but all LBS, ranging from *Google Maps* to *Foursquare*, are surveillance strategies. For these applications to work, they have to know where people are. Just as location is not inherently private, surveillance is not inherently negative. LBA can, for example, provide useful information and offers to consumers; however, issues arise when consumers have little control over how their location is being shared, which is similar to what happens in many online contexts. Scholars have persuasively written that privacy has sharply eroded online, leading to a current situation where companies collect information about almost everything and then sell that information to other companies (Solove, 2004, 2008; Turow, 2003). With the increasing number of privacy agreements people sign up on a daily basis, people have little chance of understanding how their information is being handled. The same situation occurs with the privacy policies of many LBS.

There is little in the privacy policies of mobile phone companies and services such as *Loopt* or *Yelp* that discusses how they handle the location information they share with third parties. There is also little in many of these privacy policies that identify who these third parties are. For example, Verizon's privacy policy as of November, 2010 states the following:

> Verizon Wireless understands that advances in wireless technology, especially the growing availability of location-based services, bring new concerns about how customer information is used and shared. We provide our wireless customers with clear notice regarding how these types of services work and require that they make the choice about whether specific location-tracking features available on their phones are turned on when using their wireless phones. Customers are given the opportunity to choose where and when to turn specific location-based services on and off.

This policy relies on a rhetoric of consumer choice that runs through the vast majority of advertising literature about LBA (Media advertising guidelines, 2009).

It relies on the opt-in argument that posits that consumers have control over their private information because they opt in to a service and can read the privacy policy to see what happens to their information. However, opt-in policies are often criticized for being difficult to read and unclear. Furthermore, most people do not read privacy policies or terms of use and therefore have little understanding of what companies can do with their personal information (Turow, 2003). Verizon's policy, for example, says very little about how long location information is stored or to whom it is provided. Instead, Verizon follows the above statement with a link that says the reader can click to see more information about LBS privacy. When users click on the link, however, they are taken to Verizon's home page, which is selling its VZ Navigator service. There is nothing on that page that discusses privacy in any more depth.

LBA is still in its infancy, and mobile advertising still does not represent a substantial amount of advertising budgets. Some analysts, however, suggest that this situation will likely change, which will have implications for locational privacy. For example, a *Wall Street Journal* study found that many apps transmit users' location without their consent, and one application developer claimed that he could make five times as much advertising revenue if he was willing to share users' location (Thurm and Kane, 2010). Also, Google and Apple decisively entered the LBA field in 2010, with Google purchasing the mobile advertising firm AdMob for $750 million, and Apple purchasing Quattro Wireless for $275 million and then closing it down to open its own iAd service.

It remains to be seen how LBA plans will affect users' privacy and privacy perceptions. Without clear privacy policies in place that explain exactly what will happen with users' location information, privacy concerns will become more pronounced. The thought of an individual walking down the street and being reminded on her phone that a store she likes knows where she is seems unnerving, but there is far too much potential in that scenario for companies not to try to grow LBA models, and some will certainly be useful and well received by specific consumer bases. Privacy issues ultimately boil down to control. Will users control when and with what companies they share their location information? Or will GPS just become another technology providing personal information to an untold number of databases? It is too early to know at this point. However, one thing that we do know is that, in additional to top-down surveillance, issues of collateral surveillance are very prominent with the use of location-based social networks.

Collateral Surveillance

Collateral surveillance refers to individuals having awareness of each other's locations and differs from top-down surveillance. With collateral surveillance, privacy concerns are horizontal rather than vertical and center around other people rather than institutions. In the media discourses we analyzed (de Souza e Silva &

Frith, 2010a), these concerns are expressed most frequently in the context of one specific type of LBS: location-based social networks (LBSNs). However, as we will see in the next chapter, forms of collateral surveillance are also evident in GPS-phones used for child tracking and GPS tracking systems for the elderly, travelers, and parolees.

Collateral surveillance has been around long before GPS. Jane Jacobs (1961) argues that a certain level of collateral surveillance is necessary for safe city neighborhoods. Mobile technologies have also raised issues of collateral surveillance. As we have seen in Chapter 2, the portable camera raised important privacy and surveillance issues. Taking people's pictures without their consent led to new privacy concerns, which were addressed in Warren and Brandeis' (1890) "The right to privacy." Besides leading to more substantial conceptualization of privacy, the collateral surveillance promoted by portable cameras also served to change how people thought about privacy itself. Something similar is happening now with location-aware mobile technologies.

Some location-based services, such as LBSNs work by letting people follow the location of their friends. There are two main methods by which LBSN users can control the contexts in which they share location information. The first involves limiting the number of friends with whom they share their location. Unlike *Facebook*, where some people might have five hundred friends, individuals using LBSNs are more picky about accepting new friends, a tendency which is reflected in research on the mobile social network *Dodgeball* (Humphreys, 2007, 2010). The other method is to only check in to places that are considered "public," and thus not harmful for personal privacy. A. J. Brush, John Krumm and James Scott (2010) found that although people are willing to share their location information to use LBS, many are not willing to share it when they are in or around their homes. On the other hand, there are places, such as restaurants or plazas, in which some people do not care about disclosing their location to a wide social network.

The problem with collateral surveillance and privacy is similar to what happens with the forms of "top-down" surveillance we described above: Many users have little awareness of whom actually has access to their locational information. This happens partly because of obscure privacy policies that hide the sharing of location information with third parties, but also partly because of some users' lack of understanding about how these applications work. For example, take the case of the LBSN *Whrll*. *Whrll*'s website, before it was acquired by *Groupon* in April 2011, used to show in real time on a map the locations where users were checking in. Users did have the option to choose the privacy settings of their check-ins (e.g., by limiting them to "only friends"), but the problem was that they were required to take this action for every single check-in—and the default was "public." Furthermore, by clicking on any user, anybody on the Web—even non-*Whrll* users—could have access not only to users' current location but also to their location history, which showed the most recent places they visited. The website

IcanStalkYou seeks to expose some of these issues by providing a list of all the most recent tweets that include links to geotagged photos. Many people do not know that their mobile phones include location information (in the form of latitude/longitude coordinates) in the pictures they take by default. When these pictures are posted to *Twitter,* people do not realize that they are also sharing their location with unknown users. *IcanStalkYou* shows that many people do not understand with whom they are sharing their location.

Maybe the most notable attempt to raise people's awareness about the consequences of publicly sharing their location was the website *PleaseRobMe.* *PleaseRobMe* worked by publishing public tweets from *Foursquare* users. *Foursquare* users have the option to publish their check-ins to *Twitter.* Doing so can be problematic because while *Foursquare* tends to consist of private, smaller social networks, *Twitter* is mostly public and most tweets are fully searchable and viewable by everyone. The public nature of *Twitter* is what makes it such a valuable source of real-time information, but it raises interesting privacy issues when combined with the more private locational information shared on *Foursquare.* *PleaseRobMe* exposed that problem by showing all the *Foursquare* users who had recently checked in outside their homes. The implication was that by making their location so glaringly public, people were opening themselves up to all kinds of privacy invasions. The main issue the website highlighted was that, echoing boyd (2008), location information is not binary. It is not private or public. In one context— *Foursquare*—it may be perfectly fine to broadcast location information to 30 *Foursquare* friends. In another context—*Twitter*—people may face a serious loss of privacy by sharing their location with anyone who knows how to perform a *Twitter* search.

Ultimately, collateral surveillance issues depend on how and where people share location information. Users do not automatically sacrifice their privacy anytime they share their location via a location-aware technology (Consolvo et al., 2005). When location information is unwillingly or unknowingly shared, however, people experience an invasion of privacy. This is, according to Adam Greenfield (2006), one of the problems with location- and context-aware technologies. If people understand and are able to control how they are sharing location information, and what kind of location they are sharing, then personal privacy is likely not sacrificed. However, as examples such as *Whrrl* and *PleaseRobMe* show, control can sometimes be difficult to maintain.

Reconceptualizing Privacy within Location-Aware Interfaces

To understand locational privacy today, it is helpful to look back to the late 1980s and early 1990s when Mark Weiser and his colleagues at the XEROX Palo Alto Research Center (PARC) envisioned a new paradigm for interacting with computers which they called ubiquitous computing (Weiser, Gold, & Brown,

1999). The ubiquitous computing paradigm emphasized that computing needed to leave the desktop and move out into everyday life through a network of "smart" objects—which was very different from how people imagined "computers" in the 1990s. At a time where connecting to the Internet required a desktop computer interfaced via a screen, a keyboard, and a mouse, the idea of networked interactions via "computers" spread throughout the office environment, in the form of sensors, context-aware technologies, RFID tags, and mobile devices seemed unlikely. But most importantly to our discussion, it seemed scary. Weiser, Gold, and Brown noted that as soon as their first prototypes started being tested at their lab, fears of invasion of privacy through unwanted top-down surveillance began popping up in newspapers that featured headlines such as "Big Brother Comes to the Office." However, as Weiser, Gold and Brown wisely noted, the perceived problem with context-aware technologies, "often couched in terms of privacy, is really one of control" (p. 694). If users feel in control over their location information, these technologies are not perceived as a threat to personal privacy. As such, Weiser, Gold and Brown's argument has been repeatedly used to frame issues of privacy as control in the context of digital and location-aware technologies (boyd, 2008; Perusco & Michael, 2007; Solove, 2004).

The problem with locational privacy is that people often have little control over how their locational information is shared and there are potentially serious consequences. Solove (2008) argues that there are three major problems with the current privacy landscape in regards to informational privacy: transparency, exclusion, and aggregation. All three of these concerns are related to the idea of control and apply to locational privacy. For example, the lack of transparency is pronounced in the current locational privacy landscape. The privacy policies of popular LBS and mobile phone carriers rarely delineate if they share location information with third parties, how they share the information, or if location information is stored. Privacy policies also rarely explain if companies are willing to work with the government to track users or turn over location history. The opacity of these policies gives little respect to the consumer. Just like with online services, users agree to privacy policies that are nearly incomprehensible. This lack of transparency leads to a feeling of lack of control over the technology, as many users do not understand the applications' privacy policies and therefore cannot make informed decisions about how to protect their privacy (Sadeh et al., 2009). Individuals should have the right to control who can access their information, and it is when that control is taken away that privacy issues arise (boyd, 2010; Solove, 2004; Weiser et al., 1999).

Exclusion and aggregation are also serious, interrelated concerns. According to Solove (2007), "exclusion is the problem caused when people are prevented from having knowledge about how their information is being used, as well as barred from being able to access and correct errors in that data" (pp. 766–767). As companies collect more and more data to build increasingly robust profiles, people have little recourse to access what information has been collected or whether that information is correct. The flow of information is almost entirely uni-directional, with companies collecting location information but not providing information to users about what has been collected. Exclusion generates and is generated by a power imbalance between individuals and the government, or individuals and corporations, and it relates closely to aggregation. Increasingly, companies aggregate data, that is, they combine small portions of innocuous data, such as one's zip code, shopping habits, and location history. This information is not harmful by itself and often not considered private, but when combined and/or correlated with other information, it becomes very telling about a person (Solove, 2004). These three concerns (transparency, exclusion, and aggregation) are all related to the issue of control. Because privacy policies lack transparency and people cannot access the information that has been collected, they have little control over what is done with their own locational information. As aggregate profiles are assembled, that lack of control can become more and more obvious, further driving privacy concerns people have when using certain LBS.

What makes privacy such a complicated concept is that most things are neither strictly private nor strictly public. There are exceptions of course, and most people likely have at least a few private secrets they do not want anyone else to know. For the most part, however, privacy depends on context. Location is a prime example. Location is not completely "private" information. If someone is sitting at a coffee shop, that person's location is not inherently private. She might tell her friends where she is, and even if she does not, the other people at the coffee shop will see her.

However, as we discussed earlier, people do expect to have some degree of privacy and anonymity even when in public spaces. Just because someone sits at a coffee shop (which is, according to our definition, a public space), it does not mean that she is giving permission to others nearby to access her location history, her name, or relationship status. In most cases, people at coffee shops remain anonymous to each other. And just because she is in a public space, it does not mean that she would not mind having her location made public in other public spaces, such as an online map that can be visualized by somebody else many miles away, or by a governmental agency. So, there are different degrees of publicity related to the disclosure of location information. Just because location is not inherently private, it does not mean that it is completely public. Therefore, locational privacy needs to be understood contextually. And in the context of location-aware technologies locational privacy is intimately related to the ability to control the context in which one shares locational information.

As we have discussed throughout this book, one of the keys to understanding how people negotiate urban spaces is the idea of anonymity. We saw in Chapter 2 how anonymity could be disrupted by mobile technologies such as the portable camera, which overtly identified individuals in public spaces, and kept a record of their image. These images of people in public spaces could be used in a different context, unknown to the photographed person, and this is why privacy concerns emerged. Similarly, as we mentioned earlier in this chapter, the expected anonymity in public spaces can also be disrupted when people are identified via location-aware technologies, and are unaware of the context in which their location information is used. For example, when people walk by a store and receive a perfectly timed advertisement, that advertisement is a reminder that their location is not anonymous: Companies and mobile phone providers know where they are. LBSNs such as *Gowalla* and *Foursquare* raise similar issues because often users do not know with whom they are sharing their location. If a *Foursquare* user has ten friends in her social network, but is suddenly pinged by another random user who has access to her location history and personal information through the software, she will feel uncomfortable, to say the least.

Location-aware technologies fit within the larger analysis of public spaces and mobile technology use we have been discussing thus far. We have seen that people typically avoid interacting with strangers while in public, and they have used mobile technologies as interfaces to manage their relationship to public spaces and often to avoid strangers. As mentioned in the last chapter, people often fake talking on mobile phones in public in order to avoid the awkward situation of being approached by somebody they do not know. Location-aware technology users are no different. People are comfortable sharing their location with a restricted set of known friends, but they might feel their privacy is invaded if they are randomly approached in a bar because they revealed their location on *Foursquare*. The same awkward feeling might emerge if people receive an unrequested location-based advertisement. In these cases, people can be instantly reminded that by not understanding the context of locational information sharing, they are publicizing their location and others can find them. Conversely, there are many situations in which it is desirable to share location, whether for the convenience of a relevant locative coupon or the possible serendipitous social encounters enabled by LBSNs. Understanding these issues of locational privacy requires a nuanced understanding of the contexts in which information is shared.

As danah boyd argued in her keynote speech as the SXSW conference in March 2010, privacy still matters. People still care about privacy, but as we have discussed, the way in which people understand privacy also changes. Drawing from boyd (2008), John Sloop and Joshua Gunn (2010) outline a number of prominent instances where people posted items to *Facebook* that got them fired from their jobs. The authors suggest that this may represent users who have not adapted to the new information architecture of the *Facebook* News Feed. In other words, they do not understand that the context in which they share information on *Facebook* has shifted. We may be seeing something similar with locational privacy. As more people share their location with companies and with friends, people will have to adjust to the new information architecture of the LBS market.

Questions have also been raised about how regulators will respond to the increasing use of location information. At this relatively early stage in the development of LBS, there have already been concerns expressed by legislators about locational privacy. In the United States, for example, security experts such as Ashkan Soltani (2011) have detailed to Congress the different location collection practices of major companies such as Google and Apple. Part of this congressional focus came in response to some well-publicized incidents exposing locational privacy issues, most notably the revelation that Apple's iPhone stored people's entire location history on an unencrypted file on their phones (Chen, 2011). As we mentioned above, the *Wall Street Journal* also published a high-profile article showing that many applications have no privacy policies, and some that do have them do not follow them. Making specific policy recommendations at this early stage in technological development is difficult, and there is always the serious danger that over-regulation can hinder technological innovation. For example, if legislators had regulated the early days of the Internet, there is the chance that those regulations could have strangled growth. Until a market for LBS is fully formed and revenue streams are established, it is difficult for legislators to understand the consequences of regulations. In addition, many of these applications are available in multiple countries, which means regulation can be complicated, especially if a company has to deal with multiple national regulatory frameworks. However, even while recognizing the dangers of over-regulation, we may find that some regulation is necessary in the LBS market, rather than relying on the industry to regulate itself. Even if the regulations are as basic as demanding companies not collect location information without user consent, requiring applications to include clear privacy policies, or limiting the amount of time location can be stored on servers, an increased focus on location privacy issues by regulatory agencies may help preserve people's locational privacy as long as the regulations are not overly complicated and limiting.

The development of location-aware technologies that constantly broadcast people's locations will force people and companies to re-think how they deal with location information and how they understand locational privacy. Location information will be shared in contexts to which people are not accustomed. With the constant tracking of location, individuals are forced to negotiate a new privacy landscape that includes their locational information. Therefore, new questions need to be asked about how people negotiate locational privacy. Will perceptions of privacy be influenced by the possibility of publicly sharing location information? Will people become more comfortable about sharing their locations in different contexts? As we have seen, there are different degrees to which location might be public. Sharing location with a small group of friends in *Foursquare* is different from allowing anyone that uses *Whrrl* to see one's location, which is also different from sharing one's location via *Twitter* openly on the Web. In any case, people will increasingly be confronted with these new contexts in which they share location information; to maintain locational privacy, they will have to be able to control those contexts.

Something else to consider is to which degree social norms might shift due to the increasing public character of location enabled by these devices. As Daniel M. Sutko and Adriana de Souza e Silva (2011) ask, "If your friend walks into a café and you know they have access to your location, will you question whether their appearance is serendipitous or not? What about an acquaintance, or perhaps someone who has been asking you on a date lately?" (pp. 812–813). Similarly, if somebody in a cafe sees a friend on her *Foursquare* screen who is a block away, will it be rude not to acknowledge their presence, as it would be with a friend in the same coffee shop?

The use of location-aware technologies will not only influence perceptions of locational privacy and social norms, but will also influence how people navigate public spaces. We can already see changes in the mobility patterns of users who are constantly monitored through GPS devices, such as children, parolees, and the elderly. These cases of location tracking, more than being related to issues of locational privacy, are linked to ideas of power, which is another way of understanding control in the context of location-aware mobile technologies. We address these issues in further detail in the next chapter.

Notes

1   Although collateral surveillance is not restricted to issues of privacy, this is our main concern in this chapter. In Chapter 5 we will explore the connection between collateral surveillance, personal security and control, and in Chapter 6 we address collateral surveillance and its relation to self-identity management, peer perceptions, and social convergence. As social networking scholars such as Raynes-Goldie (2010) and Albrechtson (2008) have detailed, the participatory surveillance of social media can allow for new forms of identity formation, self-expression, and empowerment, and as we discuss in Chapter 6, LBSNs are no different.

2   However, as we will see in the next chapter, this cost-benefit approach is not always fair, because people are often unaware of the kinds of personal information that are collected about them.

3   LBSNs are also known as locative mobile social networks (LMSNs). For further details, see de Souza e Silva & Frith (2010b).

References

Albrechtson, A. (2008). Online social networking as participatory surveillance. *First Monday, 13*(3).

Andrejevic, M. (2007). Surveillance in the digital enclosure. *The Communication Review, 10*, 295–317.

AT&T. (2008). AT&T completes next-generation IP/MPLS backbone network, world's largest deployment of 40-gigabit connectivity. Retrieved November 23, 2009 from http://www.att.com/gen/press-room?cdvn=news&newsarticleid=26230&pid=4800.

Barbaro, M., & Zeller, T. (2006). A face is exposed for AOL searcher no. 4417749. *New York Times.* Retrieved from http://www.nytimes.com/2006/08/09/technology/09aol.html.

Beckett, K. L., & Shaffer, D. W. (2005). Augmented by reality: The pedagogical praxis of urban planning as a pathway to ecological thinking. *The Journal of Educational Computing Research, 33*(1), 31–52.

Beirne, M., Ramsay, H., & Panteli, A. (1998). Participating informally: Opportunities and dilemmas in user-driven design. *Behaviour and Information Technology, 17*(7), 301–310.

Berggren, N., Elinder, M., & Jordahl, H. (2008). Trust and growth: A shaky relationship. *Empirical Economics, 35*(2), 251–274.

Blumberg, A. J., & Eckersley, P. (2009). On locational privacy and how to avoid losing it forever. Retrieved from http://www.eff.org/wp/locational-privacy.

Blumer, J. G., & Coleman, S. (2001). *Realising democracy online: A civic commons in cyberspace.* IPPR/Citizens Online.

boyd, d. (2008). Facebook's privacy trainwreck. *Convergence: The International Journal of Research into New Media Technologies, 14*(1), 13–20.

boyd, d. (2010). Opening remarks: Privacy and publicity: SXSW.

Brush, A. J. B., Krumm, J., & Scott, J. (2010). Exploring end user preferences for location obfuscation, location-based services, and the value of location. Paper presented at the UbiComp 2010.

Castells, M. (2000). *The rise of the network society.* Oxford: Blackwell.

Chen, B. X. (2011). Iphone tracks your every move, and there's a map for that. *Wired.*

Consolvo, S., Jung, J., Greenstein, B., Powledge, P., Maganis, G., & Avrahami, D. (2010). The Wi-Fi Privacy Ticker: Improving awareness & control of personal informaiton exposure on wi-fi. Paper presented at the UbiComp '10. Retrieved from http://seattle. intel-research.net/people/daniel/pubs/Consolvo_UbiComp_10.pdf.

Consolvo, S., Smith, I. E., Matthews, T., LaMarca, A., Tabert, J., & Powledge, P. (2005). Location disclosure to social relations: Why, when, & what people want to share. Paper presented at the Proceedings of the SIGCHI Conference on Human Factors in Computing Systems.

de Souza e Silva, A., & Frith, J. (2010a). Locational privacy in public spaces: Media discourses on location-aware mobile technologies. *Communication, Culture & Critique, 3*(4), 503–525.

de Souza e Silva, A., & Frith, J. (2010b). Locative mobile social networks: Mapping communication and location in urban spaces. *Mobilities, 5*(4), 485–506.

de Souza e Silva, A., & Sutko, D. M. (2009). On the social and political implications of hybrid reality gaming: An interview with Matt Adams from Blast Theory. In A. De Souza e Silva & D. Sutko (Eds.), *Digital cityscapes: Merging digital and urban playspaces.* New York: Peter Lang.

Dean, J., & Ghemawat, S. (2008). MapReduce: Simplified data processing on large clusters. *Communications of the ACM, 51*(1), 107–113.

Fortunati, L., & Cianchi, A. (2006). Fashion and technology in the presentation of self. In J. Höfflich & M. Hartmann (Eds.), *Mobile communication in everyday life: Ethnographic views, observations and reflections* (p. 203). Berlin: Frank and Timme.

Gordon, E., & de Souza e Silva, A. (2011). *Network locality: How digital networks create a culture of location.* Boston: Blackwell Publishers.

Greenfield, A. (2006). *Everyware: The dawning age of ubiquitous computing.* London: New Riders.

Haggerty, K. D., & Ericson, R. V. (2000). The surveillant assemblage. *British Journal of Sociology, 51*, 605–622.

Halavais, A. (2009). *Search engine society*. Cambridge: Polity.

Hashimoto, S., & Campbell, S. (2008). The occupation of ethereal locations: Indications of mobile data. *Critical Studies in Media Communication, 25*, 537–558.

Humphreys, L. (2007). Mobile social networks and social practice: A case study of Dodgeball. *Journal of Computer-Mediated Communication, 13*(1), article 17.

Humphreys, L. (2010). Mobile social networks and urban public space. *New Media & Society, 12*(5), 763–778.

Ito, M., Okabe, D., & Matsuda, M. (Eds.). (2005). *Personal, portable, pedestrian: Mobile phones in Japanese life*. Cambridge, MA: The MIT Press.

Jacobs, J. (1961). *The death of life of great American cities*. New York: Random House.

Kelly, K. (2006). Scan this book! *New York Times Magazine*. Retrieved from http://www. nytimes.com/2006/05/14/magazine/14publishing.html?pagewanted=1.

Kolmel, B., & Alexakis, S. (2002). Location based advertising. Paper presented at the Proceedings of the 2002 First International Conference on Mobile Business.

Latour, B. (1987). *Science in action: How to follow scientists and engineers through society*. Cambridge, MA: Harvard University Press.

Latour, B. (2005). *Reassembling the social: An Introduction to Actor-Network theory*. Oxford: Oxford University Press.

Lock, S. T., Shapiro, R. Y., & Jacobs, L. R. (1999). The impact of political debate on government trust: Reminding the public what the federal government does. *Political Behavior, 21*(3), 239–264.

Lyon, D. (1994). *The electronic eye: The rise of surveillance society*. Minneapolis: University of Minnesota Press.

Lyon, D. (2006a). The search for surveillance theories. In D. Lyon (Ed.), *Theorizing surveillance: The panopticon and beyond*. Portland, OR: Willand Publishing.

Lyon, D. (Ed.). (2006b). *Theorizing surveillance: The panopticon and beyond*. Toronto: Willan Publishing.

Media advertising guidelines. (2009). Retrieved from http://www.mmaglobal.com/ mobileadvertising.pdf.

Monmonier, M. S. (2002). *Spying with maps: Surveillance technologies and the future of privacy*. Chicago: University of Chicago Press.

Nakashima, E. (2007). Feeling betrayed: Facebook users force site to honor their privacy. *Washington Post*, November 30. Retrieved from http://www.washingtonpost. com/wp-dyn/content/article/2007/11/29/AR2007112902503.html.

Nguyen, D. H., Kobsa, A., & Hayes, G. R. (2008). An empirical investigation of concerns of everyday tracking and recording technologies. Paper presented at the Proceedings of the 10th International Conference on Ubiquitous Computing.

Oakes, C. (1998). "E911" turns cell phones into tracking devices. *Wired*. Retrieved from http://www.wired.com/science/discoveries/news/1998/01/9502.

Perusco, L., & Michael, K. (2007). Control, trust, privacy, and security: Evaluating location-based services. *IEEE Technology and Society Magazine, 26*(1), 4–16.

Poster, M. (1990). *The mode of information*. Chicago: University of Chicago Press.

Raynes-Goldie, K. (2010). Aliases, creeping, and wall cleaning: Understanding privacy in the age of Facebook. *First Monday, 15*(1).

Rothstein, B., & Uslaner, E. M. (2005). All for all: Equality, corruption, and social trust. *World Politics, 58*(1), 41–72.

Sadeh, N., Hong, J., Cranor, L., Fette, I., Kelley, P., Prabaker, M., et al. (2009). Understanding and capturing people's privacy policies in a mobile social networking application. *Personal and Ubiquitous Computing, 13*(6), 401–412.

Sloop, J., & Gunn, J. (2010). Status control: An admonition concerning the publicized privacy of social networking. *The Communication Review, 13*(4), 289–308.

Solove, D. (2001). Privacy and power: Computer databases and metaphors for information privacy. *Stanford Law Review, 53*, 1393–1462.

Solove, D. (2004). *The digital person: Technology and privacy in the information age.* New York: New York University Press.

Solove, D. (2007). "I've got nothing to hide" and other misunderstandings of privacy. *San Diego Law Review, 44*, 745–772.

Solove, D. (2008). *Understanding privacy.* Cambridge, MA: Harvard University Press.

Soltani, A. (2011). Testimony of Ashkan Soltani. Washington, DC: Congressional Hearing on Protecting Mobile Privacy: Your Smartphones, Tablets, Cell Phones and Your Privacy.

Staples, W. G. (1997). *The culture of surveillance: Discipline and social control in the United States.* New York: St. Martin's Press.

Stross, R. (2007). How to lose your job on your own time. *New York Times.* Retrieved from http://www.nytimes.com/2007/12/30/business/30digi.html?ex=1356670800&en=55ef6410d3cac28e&ei=5088&partner=rssnyt&emc=rss.

Sutko, D. M., & de Souza e Silva, A. (2011). Location aware mobile media and urban sociability. *New Media & Society, 13*(5), 807–823.

Turow, J. (2003). Online privacy: The system is broken. A report from the Annenberg public policy center of the University of Pennsylvania. Retrieved from http://www.asc.upenn.edu/usr/jturow/internet-privacy-report/36-page-turow-version-9.pdf.

Warren, P. (2009). The end of privacy? *Guardian.* Retrieved from http://www.guardian.co.uk/technology/2009/apr/02/google-privacy-mobile-phone-industry.

Warren, S., & Brandeis, L. (1890). The right to privacy. *Harvard Law Review, 4*(5).

Weiser, M., Gold, R., & Brown, J. S. (1999). The origins of ubiquitous computing research at PARC in the late 1980s. *IBM Systems Journal, 38*(4), 693–696.

Zetter, K. (2009). Feds 'pinged' sprint GPS data 8 million times over a year. *Wired.* Retrieved from http://www.wired.com/threatlevel/2009/12/gps-data/.

5

POWER IN LOCATION-AWARENESS

It is Monday morning, and Johanna is about to leave for school. She checks when the next bus is coming on her mobile phone screen and sees on the map that the bus is two blocks away. "I'd better rush!" she thinks. It's her first day of college, and she's anxious and worried, but also excited. Adding to her anxiety, she had just moved abroad to start school, and she knows very little about her new city.

While running down the stairs, she realizes that she has forgotten the GPS device her parents had given her. "I'm not missing my bus for that," she thinks, "after all, I'm no longer a high school student, and my parents don't need to constantly monitor me—they can't control me remotely." The bus arrives as soon as she gets to the bus stop. She gets in, sits down and immediately takes her iPhone out of her backpack. She clicks on the *LooptMix* icon. She's excited with the new app: "I wonder if there are other *LooptMix* users in this bus . . . Maybe some other student going to the same university?" She locates another nearby player who had just posted a status update that indicates he is heading to the same university. But right after she "checks in," her mobile phone battery dies. She panics. "How could I forget to charge it last night . . .? And now, how am I going to get to school?!" Johanna was counting on the GPS on her iPhone to track her movement through *Google Maps* and tell her where she had to get off the bus. Without it, she feels lost. She starts looking around at the street signs, but they are too hard to read, and anyway, she can't remember the name of the street on which she was supposed to get off the bus.

She starts looking around the bus. A man standing in front of her immediately catches her attention. He is looking at his phone. "He might be the other user I just saw," she thinks. He is wearing long pants, but looking down she sees that he is wearing an anklet. She immediately thinks about an article she had just read explaining a new parole program in California. That state had begun paroling sex offenders and requiring them to wear location-tracking devices. Her anxiety level goes up when she realizes that if he is indeed the other *LooptMix* user, now he has access to her location history, including her home location, and personal information through *LooptMix*. She never really cared about who could potentially have access to her location and personal information because she ignored the fact that sex offenders could also be *LooptMix* users. Suddenly she sees another young man with a backpack across the aisle, looking at his HTC phone. "Maybe *he* is the *LooptMix* user I saw on my screen."

Johanna is not the type who would normally start a conversation with somebody she doesn't know, much less approach a stranger on a bus, but this is her only chance. She gets up and walks across the aisle, sitting behind the potential user. From her perspective, she sees that he has the location-based social network opened on his mobile phone. It has to be him. "Excuse me," she tries. "I'm sorry to bother you, but are you going to the university?" He stops and looks at her. "Yes, you need help with something?" he answers. His response is a relief for her. Johanna then explains what happened and follows him to the campus. She feels comfortable again, and gains a new friend on her first day of school.

Location-aware mobile technologies both produce and sustain different forms of power and control. Parents give their children "chaperone" phones equipped with GPS in order to control where they can go. Parole officers remotely control parolees' mobility patterns in order to restrict the places they can visit (Michael, McNamee, & Michael, 2006; Shklovski, Vertesi, Troshynski, & Dourish, 2009; Troshynski, Lee, & Dourish, 2008). People use their GPS-phone mapping capabilities to help them feel familiar with their surrounding environment, leading to the feeling that they are able to "control" the chaos of urban spaces.

In Chapter 1, we saw that the metropolitan man's (*sic*) *blasé* attitude, described by Georg Simmel (1950), was a way of protecting the self from the overwhelming amount of external stimuli coming from the city. It was also a way of managing interactions with surrounding space and with other people in public spaces. Over the past decade, we have developed new and more sophisticated ways of managing these interactions through location-aware technologies. Different from other mobile technologies such as the book, the Walkman, and even the mobile phone, location-aware mobile technologies literally allow people to "filter" their environment by selecting the people and things with which they would like to interact. By helping users manage and control interactions with other people and with their surrounding environment, location-aware technologies become interfaces to public spaces, but they also become the instruments that mediate complex power relationships between users.

This chapter explores the power relationships embedded in location-aware mobile technology use in public spaces. Power is a form of control (Foucault, 1977). However, power does not simply mean individual control over others and over places. As Michel Foucault and Colin Gordon (1980) observe, power is not localized in the hands of specific people or institutions. It should rather be understood as a network, or a chain, something that circulates and moves, rather than a static thing. We follow Foucault and Gordon's approach that argues that "individuals are the vehicles of power, not its points of application" (Foucault & Gordon, 1980, p. 98). From this perspective, location-aware technologies themselves do not empower people to interact with spaces and other people; rather, these technologies are also elements of a power network that has the potential to discipline people's movements through space, and the social relationships in it. Often Foucault's (1977) description of the panopticon has been criticized for representing a form of top-down surveillance that is outdated because power in this model is supposedly easily identified and clearly situated (Solove, 2004). However, if we instead understand the panopticon as a mode of disciplining people's behaviors and interactions with others through the internalization of power relationships, then we are able to draw correlations with the contemporary use of location-aware technologies. By tracking the locations of people and things, new power networks are constructed, and these networks influence interpersonal relationships, people's patterns of mobility through the city, and people's relationships to places. These power networks, rather than being centralized, should be understood akin to Foucault and Gordon's (1980) idea of micro-powers: "A dispersed network of apparatuses without a single organizing system, centre or focus" (p. 71). Foucault had already observed that power does not come only from the State, but it "goes much further, passes through much finer channels, and is much more ambiguous, since each individual has at his own disposal a certain power, and for that very reason can also act as the vehicle for transmitting a wider power" (Foucault & Gordon, 1980, p. 71).

Foucault's major analysis of discipline and power took place in the context of enclosures, such as the hospital, the prison, and the school. More than a decade ago, however, Gilles Deleuze (1992) noted that we had already moved from Foucault's disciplinary societies to what Deleuze called the societies of control, in which power structures are not as evident but are nonetheless continuous. Deleuze discusses how mechanisms of control were leaving the institutional enclosure and moving into the spaces of our everyday life. In a statement that is surprisingly contemporary, he writes,

> The conception of a control mechanism, giving the position of any element within an open environment at any given instant (whether animal in a reserve or human in a corporation, as with an electronic collar), is not

necessarily one of science fiction. Félix Guattari has imagined a city where one would be able to leave one's apartment, one's street, one's neighborhood, thanks to one's (dividual) electronic card that raises a given barrier; but the card could just as easily be rejected on a given day or between certain hours; what counts is not the barrier but the computer that tracks each person's position—licit or illicit—and effects a universal modulation.

(Deleuze, 1992, p. 7)

What was science fiction to Deleuze is increasingly reality today. By identifying people's locations, location-aware technologies expand the forms of control and power into the spaces of our everyday lives, and therefore influence social and spatial relationships. As we have seen, mobile technologies such as the book and the Walkman influence how people interact with their surroundings (de Souza e Silva & Frith, 2010a). By reading a book, people can temporarily pay attention to the book's narrative and not to what is going on around them. By listening to music on an iPod users also experience city spaces in a different way. So, other types of mobile technologies already interfaced the relationships between users and places, as well as social relationships in these places. However, location-aware interfaces are different because these technologies are the interfaces of a network in which people, institutions and corporations have awareness of each other's locations—and this awareness is not always reciprocal, leading to situations of power asymmetries.

The use of location-aware technologies that enable forms of collateral surveillance, as we described in the previous chapter, may have implications not only for locational privacy, but also for sustaining different types of power relationships. For example, location-based social networks (LBSNs) such as *Loopt* and *Foursquare* raise important issues of power asymmetries, control, and exclusion in public spaces. While LBSNs are typically regarded positively for their ability to facilitate social encounters in public spaces, the filtered social space they enable may show only other nearby people and information that match the user's interests, leading to possibly exclusionary practices. Some people also fear having their location disclosed to unknown others due to issues of security. Other forms of collateral surveillance such as devices that track the location of family members (most commonly the elderly and children) both reveal and shift power relationships within close-knit social networks. However, power and control issues with location-aware technologies are not only related to forms of collateral surveillance. The increasing pervasiveness of location-based advertising (LBA) raises new concerns about the commodification of locations.

In this chapter, we examine the power relationships mediated by the use of location-aware mobile interfaces. We address forms of power asymmetries in situations of collateral surveillance with location-aware technologies in different interpersonal contexts: parolees and parole officers, children and parents, and friends in LBSNs. We then explore how location-aware technologies mediate the relationship between users and spaces. Important to our analysis is the awareness of how LBA is a form of both empowering individuals in public and of corporations exerting power over individuals who use location-based services (LBS). Finally, we explore how the use of location-aware interfaces, by allowing individuals to directly filter the information they access from their environment, leads to different forms of social exclusion and fragmented perceptions of public spaces. We finish by discussing how these new forms of location-based technology influence how we understand power and control in public spaces.

Collateral Surveillance and the Implications of Shared Location

As we have seen in the last chapter, locational privacy issues are intimately related to control. But even when privacy is not the main concern in interpersonal relationships mediated by location-aware technologies, collateral surveillance through the awareness of other people's locations also leads to power asymmetries and raises complex power dynamics in interpersonal relationships. In addition, location-aware collateral surveillance not only influences mobility patterns of those who are being watched, but also of those who watch.

Location-aware mobile technologies that enable collateral surveillance are not new. In the 1990s, when Mark Weiser and his colleagues at the Xerox Palo Alto Research Center (PARC) were defining the ubiquitous computing paradigm, they developed prototypes that tracked the location of co-workers in their offices to deliver context-based information. For example, the PARCTab communicated users' location to a central server and identified them "to receivers placed throughout a building, thus making it possible to keep track of the people or objects with which they interacted" (Weiser, 1999, p. 694). As we have seen in the last chapter, initially these prototypes were received skeptically by the popular press (Weisu, Gold, and Brown, 1999). Newspaper headlines such as "Big Brother comes to the office," highlighted the fear that PARC researchers were creating a technological infrastructure that would bring to life the top-down surveillance model described by George Orwell (1961) in *1984*. These prototypes painted the way for future, wide-ranging locative applications, but they did not expand past office environments, partly because of social fears of invasion of privacy, but mostly because of a lack of wireless infrastructure in public spaces. It was easy to set up a wireless network in an office space to test the prototypes, but during the 1990s, urban spaces did not have networks that could support devices such as pads and tabs. As computing moves out of the office and into the spaces of the city, these fears of top-down surveillance have increased. However, there are two problems with the Big Brother approach to surveillance: (1) the idea of top-down surveillance, which obfuscates other forms of decentralized and capillary surveillance, and (2) the strong emphasis on privacy concerns, which obfuscates equally relevant issues of power and control.

As we have seen in the last chapter, although many privacy concerns around location-aware technologies still rely on a rhetoric of top-down surveillance, there is an increasing number of location-aware devices that are used for collateral surveillance, enabling people to visualize the location of others. Framing the use of location-aware technologies as a process of collateral surveillance is increasingly relevant for understanding coordination in interpersonal relationships, as coordination shifts from micro-coordination with mobile phones (Ling & Yttri, 2002) to location-aware coordination (Sutko & de Souza e Silva, 2011). While micro-coordination is based on a series of short calls and text messages between two people, location-aware coordination is enabled by the visualization of people's locations on a mobile or online map. As pointed out by Sutko and de Souza e Silva (2011), micro-coordination is a more direct form of communication than location-sharing: One has to actively dial a number or type a text message instead of just following somebody else's location on a map. These forms of less direct communication and coordination result in more subtle forms of surveillance, which nonetheless reveal complex power dynamics in interpersonal relationships. For example, in situations of micro-coordination, parents keep track of their children's whereabouts by calling their mobile phones many times a day (Ling & Yttri, 2002; Palen & Hughes, 2007). In these cases, children have to stop what they are doing to answer the mobile phone, but monitoring only happens during specific times.

With location-aware technologies, monitoring becomes less intrusive (no need for a phone call), but nevertheless constant (Boesen, Rode, & Mancini, 2010; *Economist*, 2006). However, children feel they are more "free," because they do not get calls from their parents all the time. In some ways, children are empowered because they no longer need to frequently call their parents to disclose their whereabouts. Boesen et al. (2010) studied four households that used tracking technologies among family members. They found that children feel safer going to places where they would normally not go, or staying late at a friend's home because they know their parents are watching them. However, they *know* that they are being constantly watched, and this awareness of constant surveillance does influence children's range of mobility. For example, children will tend to avoid places that are seen as "problematic" by their parents (Boesen et al., 2010). Here the internalization of surveillance, as discussed in the case of the panopticon, is evidenced by a change in the children's mobility patterns. Parents no longer need to be constantly calling. They do not even need to be constantly monitoring their children's whereabouts. It is enough that the children know they *might* be tracked. However, child-tracking technologies do not only empower parents as agents of surveillance of their children's behavior, but also empower children, as they use the same technologies to subvert the act of surveillance, exhibiting the increasingly complex power dynamics of location sharing. For example, some children in the Boesen et al. study purposely "forget" their phones in a "safe" place because they do not want their parents asking questions about their location: "In one instance, a parent discovered that her child left her phone at her friend's house while she went to a party in order to stay out later than she was normally allowed" (p. 70). Furthermore, it is not only children who are constantly monitored. Parents also feel the need to be constantly checking their children's locations, and become also "trapped" into this constant surveillance network. Location-aware surveillance does not happen only within parent/children relationships. As we will see below, location-based social networks also support this type of social coordination.

The second problem with the Big Brother claim, as Weiser et al. (1999) rightly noted, is that fears of location-aware technologies should be less focused on privacy itself, and more focused on issues of control. They note,

> If the computational system is invisible as well as extensive, it becomes hard to know what is controlling what, what is connected to what, where information is flowing, how it is being used, what is broken (vs. what is working correctly, but not helpfully), and what are the consequences of any given action (including simply walking into a room).
>
> (pp. 694–695).

As we argued in the last chapter, privacy issues need to be understood within a rhetoric of control, and control is intimately related to power, that is, if somebody has knowledge (Foucault & Gordon, 1980) about the technology (how it works, how it should be controlled), that knowledge will invariably generate an imbalance in power relationships mediated by that technology. As the quote above shows, power relationships in the context of location-aware technologies nail down to what (or who) is controlling what (or whom), and where information is flowing, which may lead to cases of power asymmetries. These situations of power asymmetries are better exemplified within interpersonal relationships in examples of collateral surveillance in parolee tracking, family tracking, and location-based social networks. It is important to note, however, that the power asymmetry is not always the same. Its balance shifts according to the situation, and often power (and the lack of it) flows in both directions. For example, take the case of the parolee in the beginning of the chapter. He did not choose to have his location monitored by a parole officer, and he has little control over disclosing his whereabouts. However, although it is easy to assume that the source of power in this relationship between parole officers and parolees is unidirectional, a study by Shklovski et al. (2009) shows that the situation is much more complex.

In 2008, Shklovski et al. studied the relationships between parole officers and parolees who were being tracked by GPS anklets in the state of California (United States). In their interviews, they found that although GPS tracking should be a less intrusive form of surveillance (since before the tracking system parole officers had to closely follow suspicious parolees), it actually became a much more explicit form of control for both parolees and parole officers. On the one hand, parolees are constantly reminded of their tracking because they need to charge the anklet device every 12 hours. On the other hand, instead of following a few specific parolees, parole officers now have to keep track of many more parolees and their whereabouts, which means dealing with technological problems when the GPS does not work. Every time parolees move out of their designated areas, parole officers receive a warning on their mobile phones. The problem is that the warning can be false because of errors in GPS reading, and so parole officers are required to be in a constant state of alert. This situation clearly shows that although GPS-tracking is generally correlated with increasing surveillance and control (and therefore power), many parole officers actually feel they have less control over parolees, since the data is not always accurate, and they are not able to oversee as many parolees at once—as opposed to walking around the community and doing the regular low-tech surveillance (Shklovski et al., 2009). One disadvantage of the parolee tracking noted by parole officers is that the process is very labor intensive. Michael et al. (2006) observe that U.S. parolee officers in Georgia noted "they could easily spend an hour every morning on each offender to go over the information that is retrieved by the system" (p. 5). Another issue faced by parole officers was that they have to constantly match the locations where parolees are (as transmitted by the GPS signal) with their everyday activities (going to work, buying groceries) with the help of online maps. The problem, however, is that online maps are not always up to date, which makes this task very difficult. Often, parole officers have to ask for parolees' help in order to interpret the GPS data. So, in contrast to what one could expect, parole officers described the GPS system as a less efficient form of surveillance. As the authors note, "Ironically, the GPS system is designed such that they literally are several minutes 'behind' when actively tracking someone who is trying to abscond" (Shklovski et al., 2009, p. 6).

However, regardless of the efficiency of the system, the presence of the anklet serves as a constant reminder that the parolee is being watched. Just like with Foucault's (1977) description of the panopticon, in which prisoners did not need a prison guard because they had internalized surveillance, parolees internalize surveillance by adapting their daily lives and mobility to the constant visibility of their location to the parole officers. Parolees are allowed to move within a very restricted area, and the possibility of GPS errors limits this area even more. Interestingly, as Shklovski et al. (2009) note, it is not only the parolee who is disciplined in this process, but also parole officers. By trying to understand the system and overcome the technical difficulties described above, they also reported having to adjust their lives to the new role of location tracker. As it is clear by this example, the power network that is generated here is very similar to the situation with parents tracking children, in which parents are also "disciplined" by the use of the technology.

Issues of surveillance with location tracking are clear examples of power relationships mediated through location-aware technologies in which privacy per se is not the main issue. As Shklovski et al. (2009) observe,

The question of privacy is never far from the discussion of any monitoring system such as this. But we believe that the notion of "privacy" often hides as much as it reveals, linked as it is to particular notions of exchange and cost/benefit analysis. This particular case is a useful site at which to examine these concerns precisely because in the eyes of the law, "privacy" per se is not a relevant consideration to this parole population.

(p. 7)

Nonetheless, power relationships are overtly maintained and reinforced through sharing of location. Shklovski et al. (2009) suggest that what we are witnessing with the development of these location-based systems is the "commodification of location." In these systems, location is transformed into a digital object that can be tradable and can "move about in the (electronic world) independently from the person that it describes" (p. 8). In the case of parolees and children, the commodification of location happens when their locations (represented via a GPS trace) become a tradable entity and can be exchanged for data, negotiated with parents, handed to other authorities or care-givers, reported to other parole officers, and ultimately determine the parolee's or the child's life. Location then, "begins to have power and meaning in itself" (p. 8). Shklovski et al. suggest that in the case of the relationship between parole officers and parolees, the commodification of location displaces social relationships. But in the case of parents and children, or tracking of other family members, awareness of location is actually an intrinsic element that shapes social relations.

With some types of collateral surveillance situations, such as the tracking of elderly people, children, and travelers, people mostly agree to being tracked because they are told this is for their own safety and security. As Marc Andrejevic (2003) observes, "submission to pervasive monitoring represents a form of empowerment and, increasingly, a form of security" (p. 148). So, at the same time people feel empowered and secure by the use of the technology, they also give up power and control over their location information to have it tracked by others. To date, there are few empirical studies on the actual social implications of using location-tracking devices within families. Because the wide adoption of these services is still in its infancy, most of the studies are within human-computer interaction (HCI) literature. However, rather than focusing on large-scale social uses of these technologies, these studies are generally based on hypothetical scenarios or the empirical evaluation of prototypes among a small number of people (Barkhuus & Dey, 2003; Bentley & Metcalf, 2007; Consolvo et al., 2005; Perusco & Michael, 2007). Some scholars suggest that patterns of personal location tracking within families will follow the same models of mobile phone micro-coordination (Boesen et al., 2010). Mobile phones have been used between parents, children, and care-givers for a long time to coordinate daily activities (Ling, 2004; Ling & Yttri, 1999). Additionally, mobile phones have also been used to increase children's safety and security and to simultaneously help parents to feel more in control of their children (Devitt & Roker, 2009; Palen & Hughes, 2007). In his large-scale qualitative examination of mobile phone users, Rich Ling (2004) reported that many people adopted mobile phones because they made them feel more safe. However, as we saw in our discussion of child-tracking technology, with location-tracking the coordination shifts from a more active (voice call) to a more passive (GPS) type of interaction (Sutko & de Souza e Silva, 2011), and from a fragmented type of monitoring (frequent calls many times a day) to constant and ongoing tracking (always visible on an online map).

Child-tracking services are generally marketed as acts of love between parents and children that enhance safety and security, but they, as well as other tracking technologies intended for the elderly, in fact expose and perpetuate trust issues. In a 2007 study, Laura Perusco and Katina Michael delved into the complex relationships between privacy, control, trust, and security. They suggested that the increased control (over people) enabled by location-based services inevitably leads to a decrease in trust. Take the example of Johanna at the beginning of this chapter. Her parents wished to control her whereabouts and she perceived disclosing her location as a lack of trust in her ability to discern where to go. Even more obvious issues of trust arise when people are not aware of being tracked. For example, some of the tracked children in Boesen et al.'s (2010) study did not know they were being tracked. When they found out they were being monitored without their permission, they got angry because they perceived their parents' act as a lack of trust. However, affirming that the increased use of location-aware devices leads to a decrease in trust in interpersonal relationships is too strong a technological determinist claim. Instead of having direct effect on these relationships, location-aware technologies rather expose already existing issues. For example, some of the children in Boesen et al.'s study already had trust issues with their parents, which involved giving parents incorrect or misleading information about their whereabouts. Additionally, as we have seen with the previous examples, power never flows only in one direction. As much as parents desire to increase their control over children with the use of location-aware technologies, they are also subjected to this control, by having to constantly monitor their children's locations and (possibly) take action on it. For example, if a tracked child claims she is somewhere and the GPS trace shows a different location, how should parents confront the children?

While location sharing highlights and reinforces existing power asymmetries within domestic life, it also creates new ones. As Boesen et al. (2010) observe, location-aware technologies

> can cut across various existing power dynamics in the home—be that of parents and children or elders and adult-caretakers. Based on how it is used and implemented, LBS changes the interpersonal dynamics. Here, children were seemingly brought into line with LBS; at the same time, some found new ways of misbehaving. This bad behavior was masked by [the parents] lack of knowledge of the technology.
>
> (p. 73)

Here we can clearly see that while parents monitor their children's whereabouts, the children's knowledge of the technology empowers them to also reverse this power relationship by lying about their location. These complex power networks in collateral surveillance situations are evident in the case of parents monitoring children and parole officers monitoring parolees, but they also occur in situations that would not normally be identified as cases of power asymmetries, such as participating in a location-based social network (LBSN) or playing a location-based mobile game (LBMG).

In LBSNs and LBMGs users are generally willing to share their location with others, and they normally opt in to use the service. By opting in, they allow other people in the social network, as well as the application developer and its partners, to access their location information, though as we discussed in the previous chapter people often have little understanding of who the commercial partners are and how to manage their privacy settings for location sharing. Take the case of Johanna in the beginning of the chapter who became afraid that the supposed parolee on the bus had access to her location history and personal information through the application. Although she did not specifically share her location with him, because both were participating in the same LBSN and she never cared about adjusting her privacy settings, he could potentially access her personal information shared through the application.

In LBMGs, asymmetrical power relations arise even though players generally believe they are in control over whom they share their location information with. In their ethnographic study of the Japanese LBMG *Mogi*, Christian Licoppe and Yoriko Inada (2006) describe a case in which a player claimed to be so close to another one that he or she could actually see the other player. The observed player, without knowing if the claim was true and without having access to the other player's location—except via the game map on the mobile phone—felt vulnerable. The authors frame this situation as a power asymmetry because having access to somebody else's location (when the reverse is not true) can be conceptualized as a way of exerting power over that relationship. Something similar happened in another situation in which a *Mogi* player was close enough to another female player on the game radar (the map on the cell phone screen), but refused to identify him or herself to the female player (Licoppe & Inada, 2009). The female player got scared, and immediately called two other male players to help with the situation. The players perceived the situation as a case of stalking. While users may be more picky about whom they include on their social network on LBMGs and LBSNs such as *Mogi*, *Foursquare*, and *Gowalla* they can still easily run into a similar situation when they check into a place and an acquaintance not currently checked in anywhere finds them through the service.

Because of the possibilities of asymmetrical power relations, some LBSNs provide ways for players to manage their location information through the interface of the mobile device. For example, *Latitude* users can check in to false places, hiding their true location from acquaintances. On *Foursquare*, people can check in "off the grid," meaning they still get the points for checking in but they do not broadcast their location to friends. However, by participating in an LBSN, users are partly agreeing that their location becomes public, since they will be broadcasting it to other members of their social network, possibly to advertising companies associated with the services, and occasionally to other Internet users. As we have seen in the last chapter, some LBSN sites, such as *Whrll*, even published users' check-ins in real time on their home pages along with their location history.

As the examples above demonstrate, power relations in LBSNs are intimately connected to trust and security. The *Mogi* player did not trust the other unknown player approaching her, and therefore felt vulnerable. Our character Johanna also had to blindly trust another player on the bus in order to feel safe again in her environment. Parents give chaperone phones to children in order to "trust" that they are safe and secure. What these examples show is that there are different degrees of trust, power, and control in the context of location-aware technology use. People are forced to negotiate these issues of power and control when using location-aware technologies, and it is important to understand that there are different meanings of power and control embedded in location sharing. Furthermore, as we explained earlier, it is important to understand that power is never a unidirectional flow—it is always spread like a network, affecting all sides involved in the situation interfaced by the technology. Depending on the situation, disclosing location actually increases people's feelings of security and trust. For example, checking into *Foursquare* discloses other *Foursquare* players in the surroundings, and therefore makes players feel more familiar with (and in control of) their environment. Likewise, a child or elderly person tracked by a loved one experiences a greater sense of security.

Despite these complex power networks, the majority of media outlets we analyzed in an earlier study of media discourses on location-based services (LBS) (de Souza e Silva & Frith, 2010a) mostly focused on how tracked people exchange location information for safety and security, that is, following Shklovski et al.'s (2009) criticism, they reduce the concerns surrounding location-aware technologies to a simple cost-benefit issue. This is clear in the case of the use of LBS to coordinate rescue efforts For example, Emma Clarke (2009) from CNN.com suggested, "Privacy issues aside, the service[s] could prove invaluable for people traveling along in unfamiliar areas or dangerous situations." Similarly, mobile phone companies such as Sprint[1] portray these devices as providing safety and security to both parents and children. According to this logic, users do not mind disclosing their location information when knowledge of their location is used for their security and safety. Chaperone GPS-phones and mobile phones, for example, are thus advertised as convenient ways of tracking people, such as children, the elderly, and travelers. In these examples, location-aware technologies are seen as valuable aids in both everyday life and emergency situations due to their ability to locate people and things.

This perspective is clearly present not only on media discourses, but also in HCI scholarship. For example, a study developed in the Copenhagen airport, among the largest in Europe, asserts that most people do not mind having their location tracked through their mobile phones if it helps them feel more secure inside the airport and find their gates more easily (Hansen, Alapetite, Andersen, Malmborg, & Thommesen, 2009). This is also in line with previous research that finds that users are willing to give out location information depending on their perception of the usefulness of the application offered to them (Ackerman, Cranor, & Reagle, 1999; Ackerman, Kempf, & Miki, 2003; Barkhuus & Dey, 2003). In fact, most HCI research sees the privacy problem as a cost-benefit issue (Hong et al., 2003), that is, if individuals feel the information they submit is worth the benefits of the service, they will often use the service.[2] According to this logic, if users feel they have control over their personal settings, privacy issues tend to disappear. However, as we have seen earlier, focusing on privacy and the cost/benefit issues often obscures more relevant and complex power issues as described in this section. Although all cases above include some kind of cost/benefit trade-off, they all encompass very different location-sharing situations. It is hard to generalize power and control issues in the context of location-aware technologies, but these issues also should not be hidden solely under the idea of privacy.

Top-down Surveillance and Location-based Advertising

So far we have addressed how location-aware technology use influences power relationships in collateral surveillance situations, but power issues are also present in the relationships between people and public spaces, specifically in the context of location-based advertising (LBA). The power and control issues we have discussed surrounding the use of location-aware devices in interpersonal relationships are now transferred to the interactions between users and locations themselves. In our previously mentioned media discourses study, we found that the ability to filter location information in public spaces was generally portrayed as empowering individuals to manage their surrounding space. Media outlets normally portrayed the future of location-aware technologies by describing a scenario in which users will interact with augmented reality (AR) applications and receive relevant location-based information (including advertisements) through the interface of their mobile devices. As Kim Tong-hyung (2009) from *The Korea Times* described it, they will "hold up the[ir] phone and the screen will have names or business cards floating above the heads of pedestrians, and have advertisement menus appear next to restaurants."

The scenarios described by media outlets mostly emphasize the benefits of LBS when users are able to get location-based information in the form of restaurant recommendations and location-aware coupons. For example, an article at *Total Telcom* (2009) envisioned a future in which "a restaurant or retailer sent a text message offering a free beverage or point-of-scale coupon to someone passing the physical store." According to this logic, "chances are customers would stop in for a free drink and maybe also purchase other items as well" (*Total Telcom*, 2009). Press releases and promotional material particularly praise the potential of location-based advertising. As seen from the examples above, articles generally emphasize how users can benefit from accessing contextual and relevant information at their fingertips, but they tend to ignore more profound implications of an unprecedented commodification of locations and its implications for our interactions with urban spaces.

While LBSNs such as *Foursquare* and *Gowalla* aim at identifying other people's locations, they also provide users with specialized location-based advertisements and coupons. This business model of current LBSNs (and LBS in general) highlights yet another form of commodification of location. As we mentioned in the previous section, the commodification of location happens when locations become a tradable entity and start to have power and meaning on their own. In the case of LBA, users' locations become a valuable commodity: They are collected by mobile phone providers and traded with advertising partners in exchange for money. Advertising companies, in turn, target LBS users with ads relevant to their location. As previously discussed, it is often believed that customers trade their location privacy for the benefits of using a service (what HCI literature calls a cost/benefit issue), but frequently people are not aware that they disclose their location to commercial partners. As we saw in the beginning of the last chapter, Johanna was close to a coffee shop and she received a discount coupon for that place—although she did not directly disclose her location information to the coffee shop.

The idea of LBA is not new, tracing back at least to the early 2000s (Kolmel & Alexakis, 2002). LBA has influenced the development of location-aware technologies and applications since their inception. As Paul Dourish, Ken Anderson and Dawn Nafus (2007) note, urban computing has been originally designed with two main goals in mind:

> On the application side, many systems design efforts focus on the city as a site of consumption . . . On the user side, many systems design efforts focus their attention on young, affluent city residents, with both disposable income and discretionary mobility.
>
> (p. 2)

By urban computing Dourish et al. are referring to a myriad of devices that are used in public spaces, including mobile telephony, wireless networking, embedded computing, ubiquitous environments, and also location-aware technologies. With this in mind, it is not surprising that many of the first types of LBS lacked diversity, and were focused on consumerism. For example, most LBS so far have been developed with what they call the "Where am I?"

approach. They intend to provide users access to resources in public spaces, such as finding services (e.g., the nearest gas station, coffee shop, restaurant, etc.) and people (e.g., nearby friends via *Foursquare* and *Loopt*). While the first example is clearly intended to help users consume goods in public spaces (gas, coffee), increasingly LBMGs and LBSNs are including advertising in their software as ways to both monetize their apps and to retain existing players. For example, *Foursquare* offers coupons and free goods to players depending on the number of points they have (which depends on how many times they "check in" to specific places). Mayors of locations (people who frequently "check in" to the same place) have special privileges, and may have additional offers, as the picture in Figure 5.1 shows.

Most location-based applications are offered to the user for free with the hope developers will make money through advertising (de Souza e Silva & Frith, 2010b). Consequently, with LBS not only goods and services become commodified, but also location. With location-based advertising, the ability to attach ads and coupons to specific locations imbues them with their own value and highlights them as places of consumption. Obviously locations do not have the same value for all visitors. It is important to note that visitors to the NYC bar who are not *Foursquare* players will perceive the location differently: They will not associate it with points and badges. However, even among *Foursquare* players there are some who will purposely go to the bar to compete for the mayorship, while others will only check in sporadically. Other players will use *Foursquare* for its social networking capabilities, employing the application to coordinate meetings with a group of friends, as we will see in more detail in Chapter 6. Although the meaning of locations change with the addition of location-based information, we must remember that the information attached to a location (e.g., discounts and coupons, or tips and posts) will not solely determine how users interact with and perceive that location. If we keep in mind that location-based information is part of a set of interfaces (that also includes users, location, location-aware technologies, and location-based apps), it is clear that all parts of the system influence each other, and that these influences are not unidirectional. People come to locations with previous agendas and interests, they possess different types of technologies, and they access different types of information that is attached to a location. The interaction of all these elements contributes to the always changing meaning and differently perceived value of locations.

Locations become commodified, but it is not that there were no places of consumption in urban spaces before the emergence of LBS. Since the development of metropolises and the rise of capitalism, urban spaces are primarily spaces of consumption, embedded with advertisements in the forms of billboards and storefront signs. However, adding latitude/longitude (lat/long) coordinates to advertisements that target people depending on their location makes the relationship between people, location, and advertisements even more complex because these ads are only seen by the small subset of the population using LBS.

Figure 5.1 A *Foursquare* screen offering a location-based deal in a bar in Los
Angeles. Copyright: 2011 Foursquare.

Furthermore, ads become not only embedded in location, but also personalized,
which leads to differential spaces, as we will discuss in the next section.

Paraphrasing sociologist Henri Lefebvre (1991), LBS can then be framed as
not only a means of power and control, but also a means of "production of
location." Lefebvre coined the term "production of space" to describe ways in
which spaces are not only physical containers of people and things, but also
produced by the social interactions that take place in them. For Lefebvre, the

true meaning of a space was related to its ability to absorb connections and relationships. But Lefebvre also argues that different time periods produced different spaces. As a consequence, in our capitalist society, we produce spaces of consumption. Spaces are imbued with exchange and use value because they are produced in a capitalist system. Location-aware technologies represent an extension of this logic, but they also add something new to these spaces of consumption: the ability to attach information to places and to receive place-specific information through LBA. There is no longer, then, a shared space of consumption, but several locations of consumptions, which are differently experienced by each LBS user.

These personalized offers, coupons, and ads do not only create an individual experience of space for each user, but they also shape users' interactions with and mobility through public spaces. Exploring the connection between the consumption of space and incitement of mobility, Andrejevic (2003) argues that the ability to be connected to the Internet while moving through physical spaces enables people to carry the interactive and customizable online world into public spaces. He writes, "just as Web browsers use information about surfing habits to customize content and advertising, so the development of mobile commerce (m-commerce) promises to capitalize on the real-time monitoring of the time-space paths followed by consumers" (p. 132). According to this perspective, mobility further facilitates the consumption of space (in our case, locations) and emphasizes the locations of consumption. We can see this tendency clearly happening with LBS that adopt the "Where am I?" approach and which primarily help users to find nearby restaurants, coffee shops, bars, and gas stations. Most of these apps target users with LBA from specific locations. Rather than empowering individuals, as the popular press frequently claims, the LBA logic promises customization and individualization in exchange of constant surveillance. LBA, then, is embedded into a capitalistic logic in which being monitored now produces surplus value. According to Andrejevic (2003), this surplus value is more accentuated with newer forms of mobile consumption than with traditional online consumption because "if the individual specification of desire can work in the online monitored enclosure, it might be even more productive in the physical realm where mobility has long been correlated with increased consumption" (p. 148). As we have seen in the above-mentioned examples of popular press discourses, advertising companies and the popular press alike frame the consumption of location as a means of empowering users by helping them control and manipulate the locations through which they move. However, what often happens is exactly the opposite: Users are absorbed into a capitalist surveillance structure that only empowers those who offer and produce goods. As Andrejevic (2003) puts it,

> Consumers are told that they are now, more than ever, "in the driver's seat:" That they are taking the place of the managers who once decided

for them what products would be produced for whom. This promise, of course, is a purely formal one. Consumers are free to customize the goods and services they receive—and to pay a premium for this customization—but such freedom of choice in no way amounts to control over the production process in a substantive sense.

(p. 142)

Advertising has always been a vehicle of power in capitalist societies, and as such, it shapes how people experience public spaces. With location-based advertising, the power exerted by companies who target ads is increased because of their ability to constantly surveille consumers, which enables them to target users with contextually relevant ads.

LBA is much more likely to influence users' consumption habits. A survey conducted in August 2010 by the Mobile Marketing Association (MMA) reported that about a third (29%) of consumers in the UK, France, and Germany are likely to respond to mobile advertising and from those, over a third (39%) continue on to make the purchase (Murphy, 2010). This is substantially higher than traditional web browser banner ads, to which the response rate has been lower than one percent. And from those who clicked on the ads, normally only one percent actually buys the product (Brain, 2002; Drèze & Hussherr, 2003). More recently, Jessica Van Sack (2011), writing for the *Boston Herald*, commented that "a recent study by mobile media company JiWire found more than 50 percent of mobile users want to receive location-aware ads, and 39 percent would gladly receive coupons based on their whereabouts." Marketers see great potential in LBA because they are contextual, that is, they reach consumers right where they might want a product or offer. These messages, attached to locations, become not only an intrinsic part of that location, but also a type of lens through which users interact with public spaces.

Dourish et al. (2007) note that geographer Stephen Graham "uses the term 'software-sorted geographies' to point to the ways in which software systems increasingly act as the lenses through which we encounter the world, and, in turn, their logic becomes inscribed into those spaces" (p. 5). Graham (2005) explores the connections between the emerging software-sorting technologies and the production of spaces. He addresses how the ways in which people move and maintain social relationships are embedded into systems of power. For example, these technologies might be used to dictate patterns of mobility through the city, or to allow some people to connect while excluding others. Furthermore, they might give some people access to information while others are deprived of access. Although Graham is mostly addressing urban infrastructural technologies, such as CCTV cameras, airport biometric systems, and real-time highway pricing, his point can also be applied to the use of location-aware interfaces and LBA, which can affect how people perceive and move through the city.

LBS frame users' perspectives of their environments in a very particular way. Users are told they are empowered to control the places they occupy, but they are also subjected to various forms of surveillance. We saw in the former section how the use of location-aware technologies through tracking systems influences people's mobility through public spaces, as in the cases of parolees and children: Parolees just move about a specific area in the city, because they know they are being tracked; the same happens with children who know they are being monitored by their parents. LBS use also influences how people move through spaces even when the tracking is not obvious. For example, if an individual wants to go to an unfamiliar place in the city and uses the *Google Map* app on her mobile phone, it is likely that she will follow the path suggested by Google and will consequently not be aware of other alternative paths to get to the same destination (as would happen if she had asked somebody on the streets). Similarly, when using a recommendation service such as *Yelp* to find a restaurant nearby, people will tend to choose those places that appear on the map rather than looking around and entering a place with no reviews. These examples suggest that patterns of mobility through the city and interaction with public spaces become interfaced and modified via location-aware technologies.

Foursquare is an interesting example of how someone's path through the city can be influenced by location-based advertising. The more a *Foursquare* player checks in to a location, the better the chances to receive coupons for free stuff, such as free drinks, free cosmetics, cash discounts on dinners and groceries, and free merchandise. Frequent *Foursquare* players might tend to gravitate to the same locations to receive the benefits associated with checking in. We might then ask if the application is really fulfilling its role as a social network (by connecting people), or providing the mechanisms for business and stores to profit through the commodification of location. Perhaps it is both. And it is likely that the development of LBS and LBSN are so integrated with LBA that the focus on the city as spaces of consumption is only intensified with the development of these services.

Power and *Differential Spaces*

In his study of mobile commerce, Andrejevic (2003) notes that "the consumption of space as a means of individuation is, in general, a form of consumption limited to affluent groups" (p. 136). This comes as no surprise because advertising is generally targeted at the population who can afford to buy goods and services. As we have seen in the former section, Dourish et al. (2007) also echoed this argument, when pointing out that the design of urban computing systems focused on affluent city residents, with both disposable income and discretionary mobility. As we have seen with the examples of LBA and LBSNs in the former section, the focus on urban spaces as spaces of consumption still drives the development of LBS. As the number of location-aware phones increase, and with them the popularity of LBS and LBA, one thing we question is how the use of location-aware interfaces will contribute to new forms of *differential spaces*.

David Wood and Stephen Graham (2005) identify a type of exclusion in their discussion of *differential mobility,* defined as the exclusion of the population who do not have access to the right technologies, and therefore cannot move freely. They distinguish between high mobility, pertaining to those few with easy access, and slow mobility, which includes the majority of people who have difficult, blocked access. *Differential mobility* is directly related to power: specifically, to economic power. Those who have the access to faster transportation technologies, such as airplanes, will move faster, as will those with access to electronic road pricing and biometric documents that help travelers move seamlessly across borders. Technological systems, such as electronic road pricing and iris recognition software at airports can be understood as boundaries that control different mobility speeds and privilege what they call the "kinetic elites." These boundaries are often permeable, not obviously visible, but nevertheless very efficient in sorting out different travel speeds. As Wood and Graham (2005) note, "the more subtle and successful boundaries are *imprecise*" (p. 181).

Another example of invisible boundaries used by Wood and Graham are Internet routers programmed to deliver faster Internet access to customers based on their potential for profitability. The differential speed with which websites are accessed leads to a differential perception of online space, because more commercially attractive customers will be able to access more and faster online information than those who are not as commercially viable. While online space is experienced differently depending on connection speed, the use of location-aware mobile technologies will likely contribute to a differential experience of physical space. As we have discussed elsewhere (de Souza e Silva & Frith, 2010b), location-aware interfaces will not only contribute to differential forms of mobility, but also to what we call *differential space.* With location-aware technologies, users are able to selectively visualize people, things, and information from their surroundings. Therefore, those who carry these technologies will have a radically different experience of public spaces from those who do not. For example, if Johanna did not have her iPhone, she would probably not have been able to get to school that day unless she had asked for directions on the street. She would probably also not have met the other student in the bus. In addition, we can look at touristic practices for another example of this phenomenon. Tourists equipped with GPS-phones will be able to read dynamic place-specific information that is instantly visualized on their mobile phone screens (such as the history of monuments and buildings), which others cannot access. For example, the Argon application discussed in Chapter 3 allows tourists to create their own city tour based on their browser bookmarks about that city. If they are visiting Paris for the first time, they can download Argon and build a guide for that city based on existing browser bookmarks and web information about touristic points in the city, such as the Eiffel Tower and Champs Elysées. This information is then displayed as visitors approach those locations.[3] With the increasing adoption of these interfaces, information about a location will increasingly be stored within that location (Spohrer, 1999); consequently, people who are not able to access information will experience a different type of public space.

As was the case with *differential mobility*, it is important to be aware that different experiences of spaces are not only related to software sorting techniques and location-aware interfaces. For example, one of the characteristics of big metropolises such as Rio de Janeiro is the existence of *favelas* (the Portuguese word for slum) in close geographic proximity to the wealthiest areas of the city (de Souza e Silva, Sutko, Salis, & de Souza e Silva, 2011). High- and low-income population live so close in certain areas of the city that *favela* residents can in some instances steal items from middle-class apartments by literally "fishing" them with a fishing rod (Figure 5.2). However, these people mostly live below the poverty line and suffer from a lack of education and a lack of basic services, such as proper water pipes and electricity. Although they walk through the same streets where the elites shop in expensive boutiques, they have a radically different experience of that urban space.

However, even though differential spaces already existed before the emergence of location-aware technologies, these interfaces help users to filter their surroundings and to develop a more individualized experience of public spaces. Location-aware technologies are the interfaces that allow us to attach information to places and to retrieve place-specific information. As a consequence, locations also become filtered and manipulable in ways that were not possible before. Although issues of power, control, and exclusion have always been present in social spaces, the emergence of location-aware interfaces presents new configurations and challenges. It is no longer a matter of a binary between those who have access versus those who have no access to technology, which is how the digital divide has traditionally been conceptualized. Even among those who own the technology, if they are able to filter their environment in different ways, they will also experience spaces in individualized ways. A *Foursquare* player can identify other *Foursquare* players nearby, as was the case with Johanna, but can also then ignore other people who literally do not show up on her radar. If the other college student on the bus is a serious player, he might repeatedly "check in" the same places in order to maintain his mayorship and receive coupons and discounts to which other people who are not *Foursquare* players may not have access. For example, if he is the mayor of a bar and grill restaurant, he will keep his mayorship in order to receive a free drink every day of his reign. Plus, his favorite burger will be named after him in the menu if his reign lasts at least a week. In addition, he might not feel compelled to go to other coffee shops because this could cost his mayorship.

Therefore, these technologies contribute to exclude some by helping others to filter people and things from their interactions with space. Here we can clearly see how location-aware technologies literally interface the relationships between users and public spaces. These technologies not only help users manage their connections to public spaces, as was the case with the Walkman and the

Figure 5.2 One of the busiest streets in Copacabana, Rio de Janeiro has as its background the favela Pavão-Pavãozinho, which is in close proximity to the middle-class apartment buildings. Picture taken by Paulo Arthur Villas Boas.

iPod, but also filter that space depending on what users want to interact with. And if interfaces are also filters to public spaces, the danger is that public spaces might become over-filtered and hyper-individualized.

However, as we will see in the next chapter, there is an equally compelling counter argument, which is that sameness can pull you into difference. For example, if a *Foursquare* user sees that many of her friends are checked in at a new art gallery in an unfamiliar part of Copenhagen, she might be compelled to go there because their presence implicitly endorses/recommends that place. And this endorsement works both ways—not only do people endorse places, but also places endorse people. If an LBS user sees on her mobile phone that a place nearby has many good reviews, she will be more prone to believe that the people in there are potentially ones she would like to meet. For example, a college student who opens *Yelp* on her mobile phone finds out there are seven restaurants nearby: four in the same block, and three about a mile away. However, two of the restaurants that are further away received five-star ratings, in contrast to the three-star reviews ratings from the closest restaurants. Additionally, one of the five-star restaurants received reviews mostly from college students like her, who praised the quality and the inexpensive price of the food. She will be more inclined to go to this last place, and will perhaps make new friends.

Through these examples, it is clear that these filters to public spaces will influence how people socialize with others in public spaces, but they will also contribute to reconfigure (maintain or disrupt) power relationships that are increasingly based on the awareness of each other's locations. They will contribute both to an increased commodification of locations and to more information-rich public spaces, embedded with digital data. These public spaces might be more individualized, but they will also acquire new dynamic meanings.

Issues of power, surveillance, and control are changing with the development and popularization of location-aware technologies, and new forms of power and control require new ways of thinking about technologies and their relationship to spaces. As Simmel realized, the metropolitan man cannot go back to the small city. The shift that happened with the increasing size of metropolises, such as Paris or Berlin, along with the new challenges it posed to its citizens, such as the over-stimulation of senses, was a reality and not something that could be reverted. The same is true for the shift in the meaning of power and control within location-aware technologies we are experiencing right now. What we need, however, is to understand the shift, in order to be critical of the development and design of location-based applications that will be strongly influenced by advertisement and flows of capital, and also to address how social interactions (including power relationships) and our public spaces will change and be changed by these technologies.

Notes

1   https://sfl.sprintpcs.com/finder-sprint-family/welcome.htm.
2   However, as we saw in the last chapter, often individuals are not aware of the cost (i.e., what types of information are shared with third parties) and therefore cannot make a proper cost-benefit analysis (Consolvo et al., 2010).
3   For more information about the Argon VTG browser, see http://www.youtube.com/watch?v=01S1BbeJ-ik.

References

Ackerman, M., Cranor, L., & Reagle, J. (1999). Privacy in e-commerce: Examining user scenarios and privacy preferences. Paper presented at the Proceedings of the 1st ACM Conference on Electronic Commerce.

Ackerman, L., Kempf, J., & Miki, T. (2003). *Wireless location privacy: A report on law and policy in the United States, the European Union, and Japan.* NTT DoCoMo USA Labs.

Andrejevic, M. (2003). Monitored mobility in the era of mass customization. *Space and Culture, 6*(2), 132–150.

Barkhuus, L., & Dey, A. (2003). Location-based services for mobile telephony: A study of users' privacy concerns. Paper presented at the Proceedings of the INTERACT 2003, 9th IFIP TC13 International Conference on Human-Computer Interaction.

Bentley, F. R., & Metcalf, C. J. (2007). Sharing motion information with close family and friends. Paper presented at the Proceedings of the SIGCHI Conference on Human Factors in Computing Systems.

Boesen, J., Rode, J., & Mancini, C. (2010). The domestic panopticon: Location tracking in families. Paper presented at the UbiComp 2010.

Brain, M. (2002). How web advertising works. *HowStuffWorks.com.* Retrieved October 21, 2010 from http://computer.howstuffworks.com/web-advertising.htm.

Clarke, E. (2009, March 11, 2009). In an emergency, who do you call? *CNN.com.*

Consolvo, S., Jung, J., Greenstein, B., Powledge, P., Maganis, G., & Avrahami, D. (2010). The Wi-Fi Privacy Ticker: Improving awareness & control of personal informaiton exposure on wi-fi. Paper presented at the UbiComp '10. Retrieved from http://seattle.intel-research.net/people/daniel/pubs/Consolvo_UbiComp_10.pdf.

Consolvo, S., Smith, I. E., Matthews, T., LaMarca, A., Tabert, J., & Powledge, P. (2005). Location disclosure to social relations: Why, when, & what people want to share. Paper presented at the Proceedings of the SIGCHI Conference on Human Factors in Computing Systems.

de Souza e Silva, A., & Frith, J. (2010a). Locational privacy in public spaces: Media discourses on location-aware mobile technologies. *Communication, Culture & Critique, 3*(4), 503–525.

de Souza e Silva, A., & Frith, J. (2010b). Locative mobile social networks: Mapping communication and location in urban spaces. *Mobilities, 5*(4), 485–506.

de Souza e Silva, A., Sutko, D. M., Salis, F. A., & de Souza e Silva, C. (2011). Mobile phone appropriation in the favelas of Rio de Janeiro. *New Media & Society, 13*(3), 363–374.

Deleuze, G. (1992). Postscript on the society of control. *October, 59 (Winter)*, 3–7.

Devitt, K., & Roker, D. (2009). The role of mobile phones in family communication. *Children & Society, 23*(3), 189–202.

Dourish, P., Anderson, K., & Nafus, D. (2007). Cultural mobilities: Diversity and agency in urban computing. Paper presented at the Proceedings of the IFIP Conference Human-Computer Interaction.

Drèze, X., & Hussherr, F. X. (2003). Internet advertising: Is anybody watching? *Journal of Interactive Marketing, 17*(4), 8–23.

Economist (2006) Tracking your every move. *Economist, 381*(8506), 11.

Foucault, M. (1977). *Discipline and punish* (A. Sheridan, Trans.). New York: Pantheon Books.

Foucault, M., & Gordon, C. (1980). *Power/knowledge: Selected interviews and other writings, 1972–1977.* Brighton: Harvester Press.

Graham, S. (2005). Software-sorted geographies. *Progress in Human Geography, 29*(5), 562–580.

Hansen, J., Alapetite, A., Andersen, H., Malmborg, L., & Thommesen, J. (2009). Location-based services and privacy in airports. In T. Gross, J. Gulliksen, P. Kotzé, L. Oestreicher,

P. Palanque, R. Prates, & M. Winckler (Eds.), *Human-Computer Interaction – INTERACT 2009* (pp. 168–181). Berlin, Heidelberg: Springer.

Hong, J. I., Borriello, G., Landay, J. A., McDonald, D. W., Schilit, B. N., & Tygar, J. D. (2003). Privacy and security in the location-enhanced world wide web. Paper presented at the Proceedings of the Workshop on Privacy at UbiComp.

Kolmel, B., & Alexakis, S. (2002). Location based advertising. Paper presented at the Proceedings of the 2002 First International Conference on Mobile Business.

Lefebvre, H. (1991). *The production of space*. Malden, MA: Blackwell Publishers.

Licoppe, C., & Inada, Y. (2006). Emergent uses of a multiplayer location-aware mobile game: The interactional consequences of mediated encounters. *Mobilities, 1*(1), 39–61.

Licoppe, C., & Inada, Y. (2009). Mediated co-proximity and its dangers in a location-aware community: A case of stalking. In A. de Souza e Silva, & D. M. Sutko (Eds.), *Digital cityscapes: Merging digital and urban playspaces* (pp. 100–128). New York: Peter Lang.

Ling, R. (2004). *The mobile connection: The cell phone's impact on society*. San Francisco: Morgan Kaufman.

Ling, R., & Yttri, B. (1999). Nobody sits at home and waits for the telephone to ring: Micro and hyper-coordination through the use of the mobile telephone. Paper presented at the Perpetual Reachability Conference.

Ling, R., & Yttri, B. (2002). Hyper-coordination via mobile phones in Norway. In J. Katz, & M. Aakhus (Eds.), *Perpetual contact: Mobile communication, private talk, public performance* (pp. 139–169). New York: Cambridge University Press.

Michael, K., McNamee, A., & Michael, M. (2006). The emerging ethics of humancentric GPS tracking and monitoring. Paper presented at the International Conference on Mobile Business, June 26–27.

Murphy, D. (2010). Study highlights mobile advertising's appeal. *MobileMarketing*.

Orwell, G. (1961). *1984*. New York: New American Library.

Palen, L., & Hughes, A. (2007). When home base is not a place: Parents' use of mobile telephones. *Personal and Ubiquitous Computing, 11*(5), 339–348.

Perusco, L., & Michael, K. (2007). Control, trust, privacy, and security: Evaluating location-based services. *IEEE Technology and Society Magazine, 26*(1), 4–16.

Shklovski, I., Vertesi, J., Troshynski, E., & Dourish, P. (2009). The commodification of location: Dynamics of power in location-based systems. Paper presented at the Proceedings of the 11th International Conference on Ubiquitous Computing.

Simmel, G. (1950). *The sociology of Georg Simmel* (K. Wolff, Trans.). New York: Free Press.

Solove, D. (2004). *The digital person: Technology and privacy in the information age*. New York: New York University Press.

Spohrer, J. C. (1999). Information in places. *IBM Systems Journal, 38*(4), 602.

Sutko, D. M., & de Souza e Silva, A. (2011). Location aware mobile media and urban sociability. *New Media & Society, 13*(5), 807–823.

Tong-hyung, K. (2009). Phone and beyond. *The Korea Times*, May 11. Retrieved from http://www.koreatimes.co.kr/www/news/tech/tech_view.asp?newsIdx=44748&categoryCode=133.

Total Telecom (2009). Clickatell social network customers poised to monetise demand. *Total Telecom*, May 12.

Troshynski, E., Lee, C., & Dourish, P. (2008). Accountabilities of presence: Reframing location-based systems. Paper presented at the Proceeding of the 26th Annual SIGCHI Conference on Human Factors in Computing Systems.

Van Sack, J. (2011). Companies target ads to your coordinates. *Boston Herald.com*, Sunday, March 27. Retrieved from http://www.bostonherald.com/jobfind/news/technology/view.bg?articleid=1326300&srvc=rss.

Weiser, M. (1999). The computer for the 21st century. *SIGMOBILE Mob. Comput. Commun. Rev.*, *3*(3), 3–11.

Weiser, M., Gold, R., & Brown, J. S. (1999). The origins of ubiquitous computing research at PARC in the late 1980s. *IBM Systems Journal*, *38*(4), 693–696.

Wood, D., & Graham, S. (2005). Permeable boundaries in the software-sorted society: Surveillance and the differentiation of mobility. In M. Sheller, & J. Urry (Eds.), *Mobile Technologies of the City* (p. 177). London: Routledge.

6

THE PRESENTATION OF LOCATION

On the Saturday after her first week of school, Johanna realizes she has nothing in the fridge and she needs some milk and eggs. She heads out to one of the largest grocery stores in town. On the way, she passes by the National Museum of Art and wonders if there are any events occurring there that evening. She then opens up the app *TwitRadar* on her phone to check out what people have been tweeting about the area. The first five tweets point to a website with the schedule of the museum events. They also recommend a cheese and wine opening at 7 pm. "I'll ask Mark if he'd like to join me later. These location-based tweets are really awesome," she thinks. She had met Mark on the bus heading to her first day of class, and he's the only new friend she had made so far. She felt content, thinking all those tweets and reviews around made her more familiar with the city. She had seen a lot of positive location-based tweets about the museum's functions, and she decides that many people couldn't all be wrong.

As soon as she arrives at the grocery store, she takes out her iPhone and checks into *Foursquare*. As a newbie *Foursquare* player, Johanna checks in everywhere she goes: street corner café, grocery store, university cafeteria. She then looks at the mobile phone screen to see if her new friend is nearby. "He isn't . . ." But as soon as she raises her head, she almost bumps into Mark. She's surprised: "I wasn't expecting to see you here. You're not checked in!" He looks at her with a skeptical expression in his face: "Of course I'm not checked in. Why would I want to tell anybody I'm in a grocery store?!?" "To get points?" she suggests. "Yeah, but grocery stores aren't cool. I'd rather get points for going to the Irish Pub around the corner, or to that French coffee place." She stops and thinks for a while. She also didn't want to be known by her frequent grocery store visits. "When your social network is location-based," Johanna thought, "where you are determines who you are." From then on, she just checks in places where it is worth it to publicize her location.

Location has become an important piece of personal and spatial identity construction. In traditional social networking sites (SNS) people choose to publish pictures and select information to show their friends specific aspects of themselves. Location-based social networks (LBSNs) and location-based mobile games (LBMGs) add another element to the construction of the self: location. By choosing to check in to some places and not others, LBSN participants show their social network some aspects of their lives and not others. Those locations, then, become part of how others infer qualities about them.

LBSNs and LBMGs are not only important interfaces to social relationships among peers—they are also interfaces to public spaces. By choosing to check in to some locations and not others, LBSN users also validate places. By checking in, they are telling their friends which locations are worth going to and which are not. By writing reviews about specific restaurants and attaching them to the location of the venue, location-based service (LBS) users also attribute different meanings to those places. For example, a restaurant that gets many reviews like "the food was awful, and the waiter never came to our table" will acquire a bad reputation and no longer attract new customers. These types of reviews have been available online for quite some time, but with location-aware mobile interfaces, people are able to access them while on the move, and a bad review might literally mean a change in route. Interfaced through location-aware mobile interfaces, individuals can access other people's interpretations of those locations, through reviews, as stated above, or tips left in those locations, such as "great place to pick up art supplies for the kids," or "they ran out of refillable cups. Don't bother going to the store." As a result, people increasingly interact with digital information that has become part of locations. A user equipped with a GPS-enabled cell phone in Times Square who opens the application *WikiMe* on her cell phone is able to read *Wikipedia* articles about Times Square. As we have seen in the previous chapter, she can then have a personalized experience of that public space by manipulating and filtering of digital information. People are able to filter public spaces by making some locations more visible than others (e.g., checking in to some locations and not others), by modifying the information about a location (e.g., adding or removing pictures or sounds attached to them), or by attributing additional meaning to locations (e.g., reading or writing reviews and other texts attached to that location). Through the location-aware filter, locations are presented differently to different people. In the last chapter, we saw how these filters lead to *differential spaces* and to power asymmetries. However, the use of location-aware technologies also contributes to the construction of the very meaning of public spaces. As we will argue later, this constitutes what we call the presentation of location: *The potential to develop and access dynamic aspects of a location via location-aware technologies.*

Locations, however, are not isolated entities. They are relational and their meaning derives from connections to other locations. Consequently, locations are understood differently depending on which other locations are perceived as connected to them. For example, imagine two people having a conversation, and one telling the other about her recent trip to Europe, during which she visited the Eiffel Tower in Paris and the Coliseum in Rome, respectively. For the listener, the Eiffel Tower and the Coliseum may have been places that shared no association, but from then on they may be connected in her mind. This connection can also happen through an LBSN interface. Most LBSNs such as *Foursquare, Loopt, Brightkite, Latitude,* and *Whrll* record users' location history. By accessing a user's profile who makes that history available, one is able to see a list of all the previous locations a user has checked in since adding the application. This list, rather than only displaying isolated locations, can be seen as a particular way of influencing people's physical mobility—a user's personal mode of reading the city. Furthermore, if a user checks in to a particular location, that is a way of saying, "Hey, I'm here. Is there anybody around?" The user might intend to gather friends in a specific location, which, as a consequence, influences their way of moving through the city.

Sociability in an LBSN requires users to be mobile in urban spaces. By going from location to location, individuals are able to "read" urban spaces in new and unprecedented ways. With LBS, however, users are not only able to read spaces— they "write" spaces as well. Instead of only following someone else's spatial trail, people can choose to create trails for others to follow. By attaching meaning (in the form of text, images, etc.) to locations, people connect formerly unrelated locations. Either through LBMG play or by participating in LBSNs, people are able to create narratives as paths in public spaces.

In this chapter, we examine how location-aware interfaces enable new ways of self-presentation in which location becomes paramount to defining one's social interactions. Then we look at ways in which LBSNs and LBS allow users to personalize public spaces by attributing different meanings to locations. Finally, we argue that by following a location-based game narrative or looking at somebody's location history, individuals create their own narratives of urban spaces. Ultimately, this chapter is also about two forms of control: the ability to control the presentation of self through location, and the ability to control spaces through personalization and narrative construction.

Location as the Presentation of Self

People have always tried to control how they present themselves to others. When we go out on the streets and meet somebody, we not only infer qualities about that person based on what we can see, but we also choose to show only certain aspects of ourselves. This is what Erving Goffman (1990) called "the presentation of self." According to Rich Ling (2004), each of us fabricate our own presentation of self in order to "provide others with cues and symbols that help them place us in some context" (p. 105). Depending on the context, people present themselves in different ways. Goffman wrote about the presentation of self in everyday life as an act of performance. Each different context was described as a distinctive "stage." For example, a waiter in a restaurant will act differently depending on if he is serving clients in the main hall (front stage) or talking to the chef in the kitchen (back stage). Regardless of where we are, we are always presenting certain aspects of ourselves and hiding others. Sometimes we are at the front stage, sometimes at the back stage, but we present ourselves in different ways depending on who and what is around us. That is why for Goffman the presentation of self is not an individual act, but always based on social interaction (boyd & Marwick, 2010; Oksman & Turtiainen, 2004).

Our presentation of self while in public relies on multiple aspects that give others a strong clue about who we are, including gestures, clothing, and tone of voice. These, and other aspects of one's presentation of self rely on the body as social marker, a reliance that was complicated with the popularization of the Internet in the late 1980s and 1990s. With the Internet, people had the opportunity to represent themselves in online social spaces through avatars—digital representations of the self. The markers of the physical body that play an important role in both constructing one's own presentation of self and reading others' were lost with the turn to avatars. The possibility of presenting oneself to others via text and graphics led many to mistakenly believe that identity could be disconnected from the material body. For that reason, the presentation of self online was frequently praised because it would supposedly give people the freedom to be whoever they wanted to be (Benedikt, 2000; Dant, 2004; Lessig, 2000; Turkle, 1984, 1995; Wertheim, 1999).

This perspective permeated much of the scholarly and popular mindset in the early history of the Web. In his book *Code and Other Laws of Cyberspace*, Lawrence Lessig (2000) used the example of the deaf, the blind, and the ugly. To show the benefits of the loss of the body as social marker. In the physical world, he suggested, these people face enormous constraints on their ability to communicate that jeopardize their social interactions. When online they could use interfaces that helped them "speak" by typing, read with special software, and manipulate their appearance through avatars. Similarly, Sherry Turkle, in *The Second Self* (1984) and *Life on the Screen* (1995), extensively explored how people presented themselves in textual virtual worlds. Through interviews with multi-user domain (MUD) users, she found that individuals chose to act out certain aspects of themselves depending on which context and with whom they were interacting. This possibility of cycling through multiple self-presentations has been often interpreted as a sign of a decentered self (Nakamura, 2007; Turkle, 1984, 1995), which Lisa Nakamura (2007) framed as "identity tourism." However, the decentered self is actually nothing more than a consequence of the increased ability to control one's self-presentation enabled by online tools.

Following Goffman (1990), it is not that individuals have fragmented identities, but that they actually desire to show certain aspects of themselves and hide others. Goffman makes a distinction between expressions "given" and "given off." The former are aspects of the self over which the individual has control, and that he or she consciously decides to share with others, while the latter refers to those unintentional messages that individuals unwillingly disclose and that sometimes tell much more about who we are. The difference with online interactions, according to Judith Donath (1997) is that, "control over one's self-presentation is greater" (p. 38). In other words, the expressions "given" begin to dominate the expressions "given off."

Until the late 1990s, self-presentation online was constituted by the design of one's personal home page, the words chosen for the construction of one's textual avatar, and the color of hair, skin, and types of clothes of one's graphic avatar. In the early 2000s, however, the emergence of SNS enabled a different mode of self-presentation. To participate in an SNS, people need to create a profile, which is generally composed of one's name, age, relationship status, favorite quotes, pictures, and a network of friends. In addition to an avatar or a home page, one's friends, personal information, and pictures become part of the presentation of self.

While more options allow individuals to present more of themselves online, they can also result in less control over the presentation of self. As we have seen in Chapter 4, the increasing number of parameters people need to control to construct their online identity has often led to concerns about invasions of personal privacy. Following Goffman, people want to show certain aspects of themselves to some people and other aspects to other people. Although the expressions "given off" have always been an unintentional part of the presentation of self, in SNS this issue becomes more problematic. Suppose a college student creates a profile on *Facebook* and decides to upload pictures of last night's party where she was drunk. The pictures are intended to show some aspects of her self to a select group of friends, who (in Goffman's terms) occupy the back stage. But let's assume she forgets to choose the privacy setting that hides the pictures from another list of not-so-close friends, which includes her boss at her internship. That aspect of her self is then brought to the front stage—a context in which it does not belong. In the early days of the Web, with MUDs and USENET groups, the presentation of self online was often praised for allowing users greater control over the aspects of themselves they chose to disclose to others. However, the major concern in the recent past, with the popularization of Web 2.0 tools including SNS, has been the lack of control over the disclosure of personal information (Palfrey & Gasser, 2008; Solove, 2004). When social networks are brought into physical spaces through LBSNs and LBMGs, a new element is added to the myriad of possibilities and risks of self-presentation: location.

LBSNs complicate the presentation of self mainly because they combine traditional ways of self-presentation (via face-to-face contact) and online self-presentation. They complicate co-presence because many applications are built on the idea of meeting up with friends. For example, *Loopt*'s homepage states, "Now you can see where your *Facebook* and *Loopt* friends are on the map and let *Loopt* alert you when any of your friends are nearby so you can get together" (Loopt, 2010). Checking in, however, does not only facilitate co-presence or scoring points: check-ins also constitute a part of the presentation of self to one's social network. For example, in the story that began this chapter, Johanna decided to be more selective about the locations at which she checked in. While LBSNs have been marketed with the idea that people would like to meet others in urban spaces, many LBSN users actually check in just to broadcast their location and present themselves to their social network—even if there is no one nearby and no possibility of face-to-face meeting.

In her study of social interactions in the mobile social network *Dodgeball*, Humphreys (2007) noted that *Dodgeball* participants perceived this type of check-in as "showing off." As she notes, by checking in just to say "I'm here," *Dodgeball* members aligned themselves with particular locations and appropriated these locations as aspects of themselves (Humphreys, 2007). If locations "brand" themselves well enough (e.g., by giving out location-based coupons and discounts), more users are likely to incorporate them in their self-presentation. The commodification of location, as we discussed in the last chapter, and self-presentation through locations are intimately connected in the use of location-aware mobile interfaces. Of course, both branding and checking in affect more than just the LBSN user—they both affect how a location is given meaning through the potential to upload and access dynamic location-based information that is attached to that location via location-aware devices. We call this the *presentation of location*.

The Presentation of Location

Locations are important aspects of people's identity, but locations also have identities of their own that are formed through a combination of factors. Take the case of places. Geographical position is a relevant factor that partially forms the identity and meaning of a place. Additionally, the communities that inhabit a place also contribute to its identity (Featherstone, 1993), but it is also true that identities of places are never developed solely by their geographical position and their own people (Massey, 1994, 2005). Places are developed in relation to other people, other cultures, and other places—established by mutual difference. We have seen that the presentation of self is a relational social process because the expressions given off by people depend on who and what is around them. The same happens with places. Cities and settlements on the paths of trading routes have always experienced this relationality; their identity was constituted primarily by how their citizens saw themselves as different from others. Exchange of goods, communication among diverse peoples and cultures, and media representations of places have thus influenced the construction of the identities of places.

While geographical features play a role in the construction of a place's identity, places do not construct their identity on their own. People attribute meanings to places and are therefore an important element of their identity construction. These meanings are created by the elements that exist in a place, such as physical elements (e.g., buildings, public plazas, streets), social elements (e.g., the people who live in them), cultural elements (e.g., historical traditions, folklore, etc.), and discursive elements (e.g., media representations, travel guides). The interaction of these elements is what gives places their identity. As Lefebvre (1991) argued, spaces are not merely containers. They are constructed by social relationships, but those social relationships are not limited merely to the people who live in a specific place. The same happens with places.

The relational nature of place became more pronounced with the development of new transportation and communication technologies in the nineteenth century, such as the railway and telegraph, that brought places into contact (Carey, 1988; Schivelbusch, 1986; Standage, 2007). However, even after the nineteenth century, many of the world's cities and settlements were not reachable by the railway or the telegraph. In the twentieth century, however, newly developed electronic media, such as the radio and television, brought distant actors into people's private homes, contributing to a much greater awareness of what was happening outside the local village or community. As a consequence, people gained more awareness of distant places. Rather than just erasing local cultures and local identities, as many globalization scholars suggested (Friedman, 2007; Harvey, 1992; Virilio, 1995), Joshua Meyrowitz (2005) observed that these mass communication media actually helped people foster greater emotional attachments to places, through what he calls the "generalized elsewhere" (p. 23). According to Meyrowitz, the generalized elsewhere works as a mirror in which to view and judge our localities. The generalized elsewhere makes us more aware of our local places because they acquire relationality.

The popularity of the Internet, however, complicated the idea of the "generalized elsewhere" because unlike with the television and radio, individuals could directly communicate with distant others in chat rooms and other social platforms. As many mass media scholars observed (Boltanski, 1999; Meyrowitz, 1985, 2005; Scannell, 1996), television and radio were responsible for the feeling of making the distant near. But until recently, these mass communication media were used for uni-directional communication: A TV or radio station broadcasts a show to a large number of people, but the receivers have either none or very limited opportunity to give any feedback. With the telephone (and previously the telegraph) dialogic communication was possible, but the number of people who could be involved in a conversation was limited.[1] The Internet, however, enabled many-to-many synchronous communication. The possibility of connecting and socializing with distant others without the need of physical travel led many to believe that physical space would lose importance because people could now create online communities and develop identities independent of geographic location. As a consequence, the importance of physical space in the construction of sociability and identity online was often dismissed in scholarly discourse during the 1990s and early 2000s (Rheingold, 1993; Wertheim, 1999). According to this line of thinking, people could live in places without fully integrating into place-defined communities because they could create their own "community" in an online chat or virtual world. As Meyrowitz (2005) pointed out, "we can exit places psychologically without ever leaving them physically" (p. 27).

The predictions of the decreasing importance of place have not come true. Now, besides place, locations acquire increased relevance. The Internet is increasingly used to access local information that is organized by location. New interfaces such as mobile, location-aware technologies and mapping software make it clear that both places and locations are still important to the construction of people's identities and to the development of sociability. However, there's one main difference now: Our physical spaces are increasingly embedded with networked connections and digital information, which, as we have seen, shift the meaning of location. The phenomena that we observe now constitutes what Eric Gordon and Adriana de Souza e Silva call (2011) net locality. Net locality denotes a shift in the way we understand the Internet, in which location becomes the organizing logic of our networked interactions. As we have seen throughout this book, our location increasingly determines the types of information we access (e.g., search results on *Google Maps*, ability to download location-based information, types of *Wikipedia* articles), and as a consequence, it is no longer possible to address digital and physical spaces as separated and disconnected from each other. One no longer "enters" the Internet—it is all around us.

Net locality includes the social practices developed with the use of location-aware technologies. Net localities comprise more than only location-aware technology users. They include not only all people and things that are physically nearby, but also people and things remotely connected to those spaces. They include not only people and things, but also information attached to locations. As a result, our social spaces are no longer only constituted of face-to-face communication. As we have seen in Chapter 3, it was frequently believed that mobile technologies and remote connections would disconnect people from the interactions with their surrounding environment (Gergen, 2002, 2010; Puro, 2002), but it is no longer possible to ignore that any contemporary conceptualization of location needs to take into consideration networked connections, location-based information, and remote communication.

Consequently, locations now not only acquire meanings through their relationality to other locations and the people in them using location-aware technologies, but also through the digital information embedded in them. For example, the LBS *Socialight* allows users to attach blog posts to the locations they move through. A first-time visitor to a McLean Central Park who uses *Socialight Kids Can Play* is then able to read about how others relate to that location through the posts of the other users. For example, one user writes about the park:

Simple playground that's nice for kids aged about 3 and under. The park that surrounds it is nice, with sidewalks, green areas for kicking around a ball and trails through the woods.

If she is bringing her kids, that kind of information left by other people who frequent the park will be much more valuable than a standard brochure offered by the tourist office nearby. Or, as with Johanna and the university bar, she is able to infer qualities about places via the location-based tweets she reads. In the story, the bar near the university got a bad reputation because people started tweeting about it.

Information accessed through these services is different from reading a newspaper article about a place because location-based information frequently comes from other individuals who use the same service and who may be members of the same social network. This contrasts to the journalist who works for a newspaper, who often has no personal connection with most of her readers. Even if the local newspaper journalist is a well-known person in the neighborhood, the diversity and amount of location-based information that an app can hold is much greater than what is published in a traditional paper.[2] As a consequence, people normally relate to the location-based information differently than they do to reports from a mass communication medium such as TV or newspapers. Manuel Castells, Mireia Fernández-Ardèvol, Jack Linchuan Qui, and Araba Sey (2007) note that the personal character of text messages was one of the most important factors in the successful outcome of macro-coordination cases via mobile phones, such as the downfall of President Estrada in the Philippines in 2001 and the presidential elections in South Korea in 2003. In both cases, because the invitation to protest against the president or to participate in the elections came from a known person, or at least a friend of a friend, people felt more compelled to participate. Location-based messages and reviews operate in a similar way because they come directly from other users. These people might not know each other personally, but the fact that they have all been to the same location, and used the same service creates a link among these people, which may contribute to the trustworthiness of the shared information.

In addition to their more local character, location-based messages are different because they are attached to locations. A location-based tweet about the university bar will become permanently connected to that location and available to any *Twitter* user who visits the bar. Consequently, location-based texts uploaded through LBS (in the form of blog entries, *Wikipedia* articles, tweets, or reviews) become an intrinsic part of locations and contribute to change how people interact with those locations. The messages individuals attach to locations are important because the identity of a location is always an interplay between physical, discursive, and social elements. The designers of locations have control over their physical elements but are limited when it comes to the discursive and social. For example, the university bar in the story above already had a specific design, and its owner probably advertised it in specific ways to attract a particular group of people, but then, eventually, word of mouth gets around, and a location develops recognizable character based on frequent customers reviews and testimonials. This is the *given off* impression of a location, over which the owner has little or no control (Sutko & de Souza e Silva, 2011). The identity of the university bar shifts as the discourses about that location changed to incorporate the recent robbery. As the discourses shift, through face-to-face or online discussions or LBSNs and other communication channels such as *Twitter*, the social elements of the location change as well. The social elements of the bar—who tends to go there—are interrelated to the physical and discursive elements, which are all affected by the use of an LBSN or a LBMG. For example, if someone sees that his friends are still checking into the bar, he may still choose to go there and trust that the bar is a safe place. As Jason Farman (2011) notes, LBSNs support "a sense of social proprioception: a sense of embodied integrity that is aware of the self's place as that which is always already situated in relationship to the location of others" (p. 27). In other words, the presence of other (known and unknown) people not only helps to construct the social meaning of locations, but also contributes to the individual's awareness of their own location.

While in the last section we saw that LBSN users might infer qualities about unknown people who frequent a known location, here the presence of familiar people contributes to add different meanings to a specific location. As we have defined in the introduction, locations are increasingly also constructed by the amount of dynamic location-based information that is attached to them via location-aware mobile interfaces. Locations now are part of the logic of net locality, in which our urban public (and private) spaces are now composed of both physical and digital information, of both co-present and remote people, of both people with and without digital technologies. In fact, as conceptualized by Adriana de Souza e Silva (2006), in these new types of hybrid spaces there is no longer a distinction between the physical and the digital because they both construct the fabric of public space. Locations in hybrid spaces are constantly being inscribed and re-inscribed with location-based information. Daniel M. Sutko and Adriana de Souza e Silva (2011) call this ongoing shift in the meaning and identity of places through the use of LBS and LBSN *the presentation of place* (as a play on Goffman's presentation of self). We, however, would like to move a step forward and suggest that in fact what is happening through the interface of location-aware mobile devices is the *presentation of location* (not only places), which is highlighted by the *potential to access dynamic aspects of a location via location-aware mobile technologies*.

Location-based information can add to the presentation of location through the presence of LBSN users. When LBSN users constantly check in to a location, they encourage friends to go to that location, and may attract or repel other visitors to that location. The presence of certain people in specific locations might make visitors feel more familiar with them because they feel they can trust that location.

For example, if an American tourist travels to Brazil and "sees" on her *Foursquare* screen that many other Americans like to go to a specific bar in Ipanema beach, she might feel compelled to go there.[3] Sociability in urban environments is strongly dependent on trust (Gordon & de Souza e Silva, 2011; Sutko & de Souza e Silva, 2011). Even though people may not interact with others in a public space, they still take comfort from knowing they are surrounded by others who are somewhat similar to them. As we have seen in Chapter 1, Georg Simmel (1950) noted that although people remain anonymous to each other in urban spaces, they still trust that other people are like them, and would therefore behave like they would. That is why we feel we can safely go out in a public space where we do not really "know" anybody there.

By helping users find similar "friends," LBSNs can contribute to people's feelings of familiarity with their environment: If people are able to access information about the types of people in a location, they might be able to trust those locations and feel more comfortable in them (Sutko & de Souza e Silva, 2011). This fact has been observed in early LBMGs such as *Botfighters*. In an article about the game, the *Herald Sun* (2001) reported that some *Botfighters* players claimed to have gained motivation to go to unknown parts of their city because of the game. Bjorn Idren, a Swedish player, said that,

> eventually you start to take trips to places you wouldn't go to otherwise. I found myself sitting on the Web trying to find a nice café in an unknown part of Stockholm so that me and my girlfriend could have a picnic and also destroy a certain bot.

The knowledge about other players like him in an unknown (and perhaps distrusted) part of the city contributed to his feelings of familiarity with that location. The game motivated players to go to parts of the city to which they had never been and to "trust" these places more.

A direct implication of the *presentation of location* is that users are able to filter and "control" locations in ways that were not possible before. By being able to attach information to locations and to select nearby people and information they see on their mobile phone screen, people select aspects of locations with which they would like to interact. For example, with the LBS *WikiMe* users can not only read Wikipedia articles that are displayed according to their location, but also contribute to how other users will perceive that location by writing articles. Something similar happens with aspects of LBSNs, such as *Loopt*'s "proximity alerts." According to Farman (2011),

> As mobile devices began designing their operating systems to allow for multiple applications to run simultaneously, *Loopt* expanded the way it notified users of the proximity of those in their network. By continually running in the background, this locative social network automatically

updates your position while you move and sends an alert when someone in your network is nearby.

(n.p.)

By participating in *LooptMix*, which is not restricted to known friends, users can set preferences to be alerted when anyone with similar characteristics comes in the range of their mobile phone. For example, if Johanna sets her preference to find other undergraduate international students who like pop music, her phone will let her know when people with those characteristics are nearby.

We have seen in the first three chapters of this book that other types of mobile technologies also allowed individuals to manage their interactions with people nearby and their surrounding space. Mobile phones, for example, have been extensively studied for their ability to allow people to manage time and space through acts of micro-coordination (Ito, Okabe, & Matsuda, 2005; Ling, 2004; Ling & Yttri, 2002; Rheingold, 2002). Location-aware mobile interfaces, however, allow for a much greater degree of management of space. A book, a Walkman, or an iPod enable users to engage and disengage with public spaces on their own terms. As such, they interface the relationship between people and spaces and allow individuals to control the interactions with their surroundings. But location-aware interfaces let users not only select the amount of information they absorb from the surrounding space, but also to "change" that space via the manipulation of digital information. Individuals thus become active constructors of locations and are able to personalize those locations in new ways.

The personalization of locations is closely related to the push toward increasing customization online (Frith, in press). Many online sites, especially SNS, are customizable, becoming what Eric Gordon (2010) calls "digital possessives." Digital possessives can be identified by the constant use of the personal pronoun when referring to online sites, for instance, "my Facebook page." With the increasing adoption of location-aware mobile interfaces, digital possessives migrate to physical spaces (Foresman, 1998). Through mobile interfaces, individuals "can obtain greater control in ordering the data of people and places with which they come into contact" (Gordon, 2010, p. 163). Through the filtering and personalization of digital information, locations are presented differently to different people.

Therefore, participation in LBSNs enables new social experiences of locations. By writing a location-based blog post, publishing reviews about locations, or participating in an LBMG, people are able to collectively construct locations in ways that were not possible before. These new affordances, however, affect more than the specific location being reviewed. By attaching information to locations, participating in an LBSN, or playing an LBMG, people are also able to connect disparate locations in new and unprecedented ways. Through their mobility in space, individuals can re-read and re-write urban spaces.

Urban Narratives

In his book *Consumers and Citizens*, Néstor Garcia Canclini (2001) writes,

> Walking through the city [Mexico City] is like a video clip in which diverse musics and stories are mixed . . . One alternates passing by seventeenth-century churches with nineteenth-century buildings and constructions from every decade of the twentieth, interrupted by gigantic billboards layered with models' phony bodies, new cars, and newly imported computers. Everything is dense and fragmentary . . . The city has been created by plundering images from everywhere, in any order whatsoever. Good readers of urban life must adapt themselves to the rhythm and bliss out of the ephemeral visions.
>
> I end up asking myself if we will be able to narrate the city again. Can there be stories in our cities, dominated as they are by disconnection, atomization, and insignificance?
>
> (p. 85)

Canclini points to the fragmentation of the urban landscape, both through the influences of history and globalization. The context in which Canclini was writing encompassed an increasingly globalized world, in which people feared the disappearance of national borders and the erasure of national cultures. Canclini argued that though local cultural and social characteristics were influenced by the globalization process, they nevertheless did not disappear. They instead co-exist with other influences, leading to a fragmented and disconnected urban landscape. He inquires whether, in the face of all these influences fighting for space in the city, it is possible to make sense of the urban landscape and associate apparently disconnected places. To Canclini's question, we suggest that yes, it is possible to narrate the city again. Location-aware interfaces will play an increasingly important role in helping to construct urban narratives and define trajectories through the urban landscape.

Canclini (2001) observed that narrating the city at the end of the twentieth century meant knowing that it was no longer possible to have the experience of order that Baudelaire's *flâneur* expected to find in his nineteenth-century strolls through the city. More than one hundred years after Baudelaire's *flâneur*, contemporary megacities, such as Mexico City, São Paulo, Caracas, and Mumbai are filled with so much stimuli and information that they are more "like a video clip, an effervescent montage of discontinuous images" (p. 84). Therefore, he notes, it is no longer possible to go from one place to another without being disturbed by all sorts of images, sounds and people. Finally, he adds, "it is no longer possible to imagine a story from the purview of a historical or modern center that would permit us to draw the only possible map of a compact city that has ceased to exist" (p. 85).

We agree with Canclini that it is no longer feasible, if it ever was, to draw only one map of a city. Rather, urban spaces can be represented and narrated in many different ways, producing multiple maps, each of which contains different elements and perspectives of the urban environment. Furthermore, maps now become dynamic, ever changing, and embedded with digital information that is constantly being produced and re-created. One might ask, as Valentina Nisi, Ian Oakley and Mads Haahr (2008) do, "how can stories be distributed and fragmented but remain meaningful?" (p. 4). Increasingly, location-aware mobile interfaces are important tools for constructing meaning through this fragmentation. These interfaces enable us to connect fragmented locations in the urban landscape, and by doing that, they create new forms of mobility and mapping within the city.

Reading Location

As we explained earlier in this chapter, when people tell a story that includes different places, these places become associated in the listener's mind. The production of location-based information also has the power to associate disparate locations. Either through a user's location history or through parts of narratives and texts embedded in different locations, users create associations and connections among previously disconnected locations in the city; they suggest patterns of mobility through urban spaces. Other users may follow these patterns and "read" urban spaces in ways that were not previously possible, showing how embedding location-based information leads to new ways of connecting and moving through urban spaces.

As we have seen in the last section, LBS such as *Socialight* and *Yelp* enable people to write posts and reviews about locations and imbue those locations with new meanings. While reading about specific locations, people are also able to create their own associations among locations. For example, a *Loopt* user who checks in at the neighborhood movie theater and posts "going to Central Park" as a status update creates a link between formerly unrelated locations—in this case, Central Park and the theater. This post might be read by another *Loopt* user, who in turn will associate these two places in her mind. Similarly, a group of *Foursquare* users who go barhopping, checking in at every bar they visit, will also create new associations among those (perhaps) previously unrelated locations. These connections might then encourage other *Foursquare* users to follow the other users' trajectory through the city, and will therefore represent a new way of moving through the city and "reading" its locations.

This experience of reading the city is even stronger with other LBMGs such as *Botfighters* or *Mogi*, in which players are required to actively move through the city in order to play the game (de Souza e Silva, 2009; Licoppe & Guillot, 2006; Licoppe & Inada, 2006; Sotamaa, 2002). In most videogames, players are the main character of the games, but they are represented by an avatar on the screen (Calleja,

2011; Mateas, 2001). In LBMGs, players literally embody the game's character, who moves through the city in order to complete the game's goal (de Souza e Silva, 2009). For example, when playing *Alien Revolt*, players had to embody an alien or a human as their character and engage in a role-playing adventure in order to save (or conquer) Earth (de Souza e Silva, 2008). Players also had to search for other players in public spaces in order to engage in combat. The story of the game thus becomes intrinsically connected to the players' mobility patterns through the city. Furthermore, as explained earlier, by playing the game, players will experience city spaces differently from before. What was previously just a bakery now becomes the alien group meeting point, or the neighborhood park becomes a battlefield.

According to Johanna Brewer and Paul Dourish (2008), one of the functions of networked mobile technologies is to increase the "legibility of spaces and actions—how it is they can be read and understood as conveying particular sorts of messages" (p. 971). Indeed, following a path through urban spaces can be compared to hypertextual reading. In the mid-1990s, when the World Wide Web was still in its infancy, hypertexts received increasing attention from the scholarly community as a new mode of reading, which is not to say that non-linear reading started with the WWW. Since the invention of the codex book and its popularization with the printing press, non-linear reading was possible through footnotes and tables of contents that forged semantic connections with specific sections within the main text. These connections, however, were static and unless a book was redesigned and reprinted, the order of pages remained the same. Hypertexts are different in large parts because they were not static and strictly ordered.

The structure of hypertext, by which information chunks are associated by meaning, had been originally envisaged by Vannevar Bush's design of the Memex (1945). The Memex used associative trails to connect information in a logical way and greatly influenced twentieth-century information architecture. In his design of the Memex, Bush was in fact imagining a new type of interface. He was also creating a new mode of reading that precluded a pre-established textual order. Ted Nelson labeled this new structure *hypertext* in the 1960s. The Memex was envisioned to help users filter information stored in a huge database made of microfilm. The Web, although comprising digital information, and undoubtedly larger, can also be thought of as a series of databases (Manovich, 2007; Murray, 1997). Building on that idea, Lev Manovich suggested that the Web is composed of two main characters: interfaces and databases. While databases are collections of elements, subdivided into categories, interfaces are ways of accessing databases and of rearranging their elements in a linear, human-like way. Interfaces, according to Manovich, allow users to read information in different ways. Like Manovich, many scholars emphasized this disconnection between the database (the information stored) and the interface (modes of reading) that occurred on the Web.

For example, in his book *Technologies of Intelligence*, Pierre Lévy (2004) suggests that in oral cultures narrators and storytellers had a dual role: They were the ones who transmitted and stored knowledge (information) as well as the ones who interpreted this information. They were both the "databases" and the "interfaces." When writing emerged, this connection was destroyed. Walls, clay tokens, and papyrus became the information storage devices. After the Web, the way of creating narratives was once again challenged because the order of a text was no longer determined by the author, instead being determined by the reader who decided which links to follow. As a consequence, the author of a hypertext was no longer the one who wrote a story from the beginning to the end, but the one who stored information that could be accessed ("read") in a variety of ways.[4]

Hypertextual reading is important for how we view paths through the city because with location-aware mobile interfaces and the ability to attach information to locations, urban spaces are now transformed into databases. The pieces of location-specific information receive lat/long coordinates, waiting to be activated according to the location of the users through their location-aware mobile device. As such, by accessing location-based information, individuals read the city as a hypertext. The ability to connect and associate different locations, triggered by their movement through space, imparts a new, perhaps more ordered experience of urban space (Frith, in press). Location-aware technologies then become the interfaces that enable people not only to filter the information that is attached to locations, but also to connect to them in multiple ways. The hypertextual mode of reading is thus transferred into urban spaces.

The ability to read physical spaces by connecting information as one moves through space predates the development of location-aware technologies. Artist Janet Cardiff has been developing audio walks since the early 1990s. Cardiff does not use any type of location-aware technology in her walks. Instead, a person listens to a pre-recorded tape with instructions about how to move in space. For example, her 1997 *Chiaroscuro* walk, created for an exhibition at the San Francisco Museum of Modern Art, was her first walk to be experienced inside a museum. At the beginning, the person experiencing the audio installation listened to her voice: "Push the elevator button. We'll go down to the first floor." Then a bit later, "Let's go up. Go around the corner to the front of the main stairs. Walk up the stairs. I'll walk slowly so we can stay together" (Cardiff & Bures Miller, 1997). By giving the museum visitor instructions on how to move through the space, Cardiff was allowing the listener to share her connections among the museum's different rooms. The narrative created by Cardiff works as an audio layer on top of physical space that influenced how people experienced both the museum and their personal mobility.

Cardiff's audio walks, however, are different from a location-based experience because the spatial narrative is not triggered by the location of the user. To experience Cardiff's narrative, participants have to proceed in the order dictated by the audio narration. If they do not, they can still hear the narrative but they will not be in the physical location it is describing. In these cases, there is only one possible way of connecting the narrative chunks: the one chosen by the artist. However, with the removal of the Global Positioning System (GPS) signal degradation in 2000, soon other media artists started employing GPS devices as interfaces to create location-based audio walks. One of the first was *34 North 118 West*, developed in 2002 by Jeff Knowlton, Naomi Spellman and Jeremy Height (Knowlton, Spellman, & Hight, 2002). Named after the lat/long coordinates of the city of Los Angeles, *34 North 118 West* tells a story about the railway industry in downtown Los Angeles at the turn of the century. As participants navigate the city streets, they hear the voices and activities of people who lived and worked in the area. In order to experience the narrative, participants were given a tablet PC and headphones. On a map on a tablet PC, they could see the location of hotspots. Approaching a hotspot triggered the story attached to that location. Each individual who experienced the piece followed a different narrative by walking through different hotspots visualized on the map.

In their documentation video, the artists claim that the landscape had become the interface. But if we consider, following Manovich, that interfaces are the filters that allow the linking of data in different ways, the location-aware devices become the interface and the landscape is the database. The tablet PC is the interface that allows people to filter the digital information embedded in public spaces and also allows them to connect this information in order to create different narratives. These audio walks begin to point to the hypertextual nature of paths through space, in which individuals "read" the city differently by interacting with location-based information attached to the locations they visit, and therefore constructing their personal spatial trajectories.

Steve Benford, Gabriella Giannachi, Boriana Koleva, and Tom Rodden (2009) developed the concept of interactive trajectories, defined as people's journeys through space, "shaped by narratives that are embedded into spatial, temporal and performative structures by authors" (p. 712). They, however, clearly oppose the experience of trajectory to the hypertextual mode of reading. The authors claim that while hyperlinks are "discreet but connected," trajectories are "continuous and interwoven." In their analysis, Benford et al. are referring to specific types of experiences that are normally controlled by their creators, such as their pieces *Can You See Me Now?* (Blast Theory & Mixed Reality Lab, 2001) and *Uncle Roy All Around You* (Blast Theory & Mixed Reality Lab, 2003). Because these are controlled experiences, they function similarly to audio walks and location-based sound installations, which are also clear examples of interactive trajectories, that is, participants follow pre-determined paths envisioned by the work's creator. Benford et al. suggest that these "interactive experiences enable each participant to define their own trajectory, making individual choices and following personal routes" (2009, p. 715). However, with these pre-determined paths designed by developers, the "individual choices" and personal routes are severely constrained. As we discuss below, typical interactions with LBS, LBSNs, and LBMGs resemble hypertexts more than orchestrated trajectories, because these services also enable people to define their own trajectories in public spaces, but without the orchestration provided by the artists.

The ability to create personal spatial trajectories using location-aware technologies interweave physical and digital spaces, creating a different experience of hybrid spaces. Adriana de Souza e Silva (2006) defines hybrid spaces as social spaces created by the combination of physical space with digital information and the mobility of users equipped with location-aware interfaces. For example, *Mogi* required Tokyo residents to complete a collection of virtual objects and creatures that were attached to specific locations. Christian Licoppe and Yoriko Inada (2006, 2009) found that some *Mogi* players changed their route from home to work—for example, going by bus rather than by subway—so that they could find and collect more objects. Once downloaded to their mobile phones, these objects could then be exchanged with other players. Other players decided to go out at night in groups to collect particular objects. Players constructed their own trajectory through their mobility through physical and digital spaces. LBMGs, LBSNs, and LBS allow users to link locations and "read" the city in different ways, because the locations visited acquired distinct meanings, depending on the location-based information encountered by players.

In art installations such as *34 North 118 West*, participants are able to read space by following chunks of information that somebody else created. Something similar happens when people use LBS such as *Socialight* and *Foursquare*. When using these applications, users mostly follow trails left by somebody else—the user who uploaded the *Foursquare* tip or wrote the *Socialight* post.[5] But when people start contributing to create the information that is attached to a location, they actively create the links among these locations. People are then transformed from readers into writers of urban spaces.

Writing Location

Almost all LBSNs and many LBS allow users to attach information to locations. As we discussed in the last section, embedding information in locations contributes to change their identity (the presentation of location), and it also influences new patterns of mobility through the city through the linking of locations. LBS that allow users to "write the city" can be traced back to digital annotation projects that let users create their own content instead of just reading content created by others. For example, Proboscis' *Urban Tapestries* (2002–2004) was perhaps the first mobile annotation project. Developed in partnership with collaborators from academia and industry, the project aimed at allowing users to "author" the environment around them by uploading location-specific information, such as stories, sounds, and videos. The real innovation of *Urban Tapestries* was to develop a location-aware platform that challenged the common top-down approach of location-based experiences at that time, in which users would just access information uploaded by others (Angus, 2003; Silverstone & Sujon, 2005). *Urban Tapestries* enabled users to share knowledge and create a series of "threads" that consisted of information about local resources that linked locations to people.

More recently, Anders Sundnes Løvlie (2009) developed a project named *Textopia*, which was later on made available as a mobile application in the Android marketplace. *Textopia* allowed users to walk through the city of Oslo (and now elsewhere) and listen to location-based texts via their mobile phones. However, because one of the artist's aims was to make it easy for users to also contribute their own texts to the piece, he created a wiki to serve as a database of texts. To populate the wiki, Løvlie organized a competition in which users could create their own location-based texts and submit them to the system. One of the interesting findings of this experiment was that there were three main categories of texts uploaded by users: *placetexts*, *voice sculptures*, and *stray voices*. While *placetexts* and *voice sculptures* were mostly stories and descriptions of locations (something similar to what users generally do when writing a *Socialight* post or a location-based review on *Yelp*), *stray voices* had a distinct character. According to Løvlie, these were a "series of texts which form stories that move from place to place, requiring the user to physically traverse the landscape of the story in order to traverse the text of the story (though not necessarily in linear sequence)" (2009, p. 22). *Stray voices* are types of texts that truly constitute a location-based narrative because they explicitly engage people in connecting locations, by requiring them to move through public spaces. We suggest that *stray voices* encourage specific spatial trajectories (Benford et al., 2009), but these spatial trajectories are also created by common users, rather than solely by artists. Through *stray voices*, people can create stories about the city and locations acquire different meanings depending on people's mobility through space. In addition, those who write these urban stories are actively contributing not only to add new meanings to locations (the presentation of location), but also to "write" urban spaces in a very specific way.

Not all location-based information is structured and organized as narratives and spatial trajectories. Nonetheless, they are ways of writing space. LBSNs and LBMGs show this quite well. There are different ways of engaging with LBSNs: checking in, putting up status updates, acquiring coupons, gaining points, engaging with a game narrative, and sharing location. Humphreys (2007) noted that *Dodgeball* enabled users to "catalog" their lives by recording each past location, which could later be visualized on a Google map. As we have seen, the collection of past locations is part of the LBSN user's presentation of self. Even if unintentionally, cataloging creates new connections and links among locations that now can be clearly visualized on a map of the city. However, as Canclini mentioned, it is no longer feasible to have only one coherent map of the city. Within this fragmentation, each person, with their movement, links together disconnected elements and creates their own threads and maps with their mobility through the urban space, and each set of combined locations constitutes a different way of reading and inscribing the city.

The city, then, can be compared to a palimpsest, containing different layers of text that can be read and re-read in random order. A palimpsest originally referred to a parchment that had been written upon or inscribed more than once. This practice was very common in the Middle Ages, when the parchment used for manuscripts was very expensive and it was thus necessary to "recycle" the used material. Because the act of erasing the parchment was not perfect, the previous text or texts remained partly visible. This process unintentionally created several layers of text on the same surface, generating many layers of meaning. The same happens with city spaces. For Canclini, these different levels of meanings could not form a coherent narrative. But now, by writing space, reading, and re-writing again, users of LBSNs and LBMGs do create multiple levels of meanings. Each new pattern of movement, each new location-based text does not erase the previous one, but contributes to generate new textualities and social interactions in public spaces.

Notes

1   Even in the case of a conference call or wireless radio, the number of people that can be involved in the communication channel is limited when compared to the affordances of chat rooms and virtual worlds.
2   Note here that we are contrasting LBS with traditional printed newspapers. However, with increased media convergence, many newspapers are developing mobile apps and many newspaper articles allow readers to add commentaries. However, a newspaper article that allows comments is very different from a location-based app in which the bulk of information that it contains is developed by users.
3   Obviously, this can also work the other way around, where people avoid some places because they know who is there at the moment or who has been there before.
4   It is worth noting that many modern narrative theorists argue that the meaning of texts has never been solely constructed by authors (Barthes, 1977; Eco, 1989). However, what is at stake with the current notion of hypertext is the change in the materiality of the storage device and the inscription interface: a move from the analog pages on a book, to a digital database, which is capable of storing a substantially larger amount of information. Additionally, because databases are composed of disparate items (Manovich, 2002), it is possible to connect this item in multiple ways, producing different readings of it.
5   Note that although it is possible to read a *Foursquare* tip even when one is not at the exact location where the information was written (as it is also possible to listen to a *34N118W* vignette from the authors' website), we are focusing here on the connections to locations that are experienced when people are able to read space by downloading location-specific information at the location where the information belongs. Although one can read or listen to this information remotely, only by experiencing the information while in the location it describes does it have the power to change patterns of mobility in real time.

References

Angus, A. (2003). Urban tapestries contexts. Retrieved November 1, 2009 from http://www.vimeo.com/1065977.

Barthes, R. (1977). The death of the author (1968). *Image, Music, Text*, 142–148.

Benedikt, M. (2000). Cyberspace: The first steps. In D. Bell, & B. M. Kennedy (Eds.), *The cybercultures reader* (p. 29). New York: Routledge.

Benford, S., Giannachi, G., Koleva, B., & Rodden, T. (2009). From interaction to trajectories: Designing coherent journeys through user experiences. Paper presented at the Proceedings of the 27th International Conference on Human Factors in Computing Systems.

Blast Theory, & Mixed Reality Lab (Artist). (2001). *Can you see me now?* (Hybrid reality game).

Blast Theory, & Mixed Reality Lab (Artist). (2003). *Uncle Roy all around you* (Hybrid reality game).

Boltanski, L. (1999). *Distant suffering: Morality, media and politics*. Cambridge: Cambridge University Press.

boyd, d., & Marwick, A. E. (2010). I tweet honestly, I tweet passionately: Twitter users, context collapse, and the imagined audience. *New Media & Society, xx*(x), 1–20.

Brewer, J., & Dourish, P. (2008). Storied spaces: Cultural accounts of mobility, technology, and environmental knowing. *International Journal of Human-Computer Studies, 66*(12), 963–976.

Bush, V. (1945). As we may think. *The Atlantic Monthly, 176*(1), 101–108.

Calleja, G. (2011). *In-Game: From immersion to incorporation*. Cambridge, MA: MIT Press.

Canclini, N. G. (2001). *Consumers and citizens: Globalization and multicultural conflicts*. Minneapolis: University of Minnesota Press.

Cardiff, J., & Bures Miller, G. (1997). Chiaroscuro, 1997. From http://www.cardiffmiller.com/artworks/walks/chiaroscuro.html.

Carey, J. (1988). *Communication as culture*. Boston: Unwin Hyman.

Castells, M., Fernández-Ardèvol, M., Qiu, J. L., & Sey, A. (2007). *Mobile communication and society: A global perspective*. Cambridge, MA: MIT Press.

Dant, T. (2004). The driver-car. *Theory, Culture & Society, 21*(4–5), 61–79.

de Souza e Silva, A. (2006). From cyber to hybrid: Mobile technologies as interfaces of hybrid spaces. *Space and Culture, 3*, 261–278.

de Souza e Silva, A. (2008). Alien revolt: A case-study of the first location-based mobile game in Brazil. *IEEE Technology and Society Magazine, 27*(1), 18–28.

de Souza e Silva, A. (2009). Hybrid reality and location-based gaming: Redefining mobility and game spaces in urban environments. *Simulation & Gaming, 40*(3), 404–424.

Donath, J. (1997). Inhabiting the virtual city: The design of social environments for electronic communities. Unpublished PhD Dissertation. Massachusetts Institute of Technology.

Eco, U. (1989). *The open work*. Cambridge, MA: Harvard University Press.

Farman, J. (2011). *The mobile interface of everyday life: Technology, Embodiment, and Culture*. New York: Routledge.

Featherstone, M. (1993). Global and local cultures. In J. Bird, B. Curtis, T. Putnam, G. Robertson, & L. Tickner (Eds.), *Mapping the futures: Local cultures, global change*. London, New York: Routledge.

Friedman, T. L. (2007). *The world is flat 3.0: A brief history of the 21st century*. New York: Picador.

Gergen, K. (2002). The challenge of absent presence. In J. Katz, & M. Aakhus (Eds.), *Perpetual contact: Mobile communication, private talk, public performance* (pp. 227–241). New York: Cambridge University Press.

Gergen, K. (2010). Mobile communication and the new insularity. *Interdisciplinary Journal of Technology, Culture and Education*, 5(1).

Goffman, E. (1990). *The presentation of self in everyday life* (rev ed.). New York: Doubleday.

Gordon, E. (2010). *The urban spectator: American concept cities from Kodak to Google*. Lebanon, NH: Dartmouth College.

Gordon, E., & de Souza e Silva, A. (2011). *Net locality: Why location matters in a networked world*. Boston: Blackwell Publishers.

Harvey, D. (1992). *The condition of postmodernity: An enquiry into the origins of cultural change*. Malden, MA: Blackwell.

Herald Sun (2001). Mobile killers. *Sunday Herald Sun*, July 15, p. 89.

Humphreys, L. (2007). Mobile social networks and social practice: A case study of Dodgeball. *Journal of Computer-Mediated Communication*, 13(1), article 17.

Ito, M., Okabe, D., & Matsuda, M. (Eds.). (2005). *Personal, portable, pedestrian: Mobile phones in Japanese life*. Cambridge, MA: The MIT Press.

Knowlton, J., Spellman, N., & Hight, J. (Artist). (2002). *34 North 118 West*.

Lefebvre, H. (1991). *The production of space*. Malden, MA: Blackwell Publishers.

Lessig, L. (2000). *Code and other laws of cyberspace*. New York: Basic Books.

Lévy, P. (2004). *As tecnologias da inteligência: O futuro do pensamento na era da informática*. São Paulo: Editora 34.

Licoppe, C., & Guillot, R. (2006). ICTs and the engineering of encounters: A case study of the development of a mobile game based on the geolocation of terminals. In J. Urry, & M. Sheller (Eds.), *Mobile technologies of the city* (pp. 152–163). New York: Routledge.

Licoppe, C., & Inada, Y. (2006). Emergent uses of a multiplayer location-aware mobile game: The interactional consequences of mediated encounters. *Mobilities*, 1(1), 39–61.

Licoppe, C., & Inada, Y. (2009). Mediated co-proximity and its dangers in a location-aware community: A case of stalking. In A. de Souza e Silva, & D. M. Sutko (Eds.), *Digital cityscapes: Merging digital and urban playspaces* (pp. 100–128). New York: Peter Lang.

Ling, R. (2004). *The mobile connection: The cell phone's impact on society*. San Francisco: Morgan Kaufman.

Ling, R., & Yttri, B. (2002). Hyper-coordination via mobile phones in Norway. In J. Katz, & M. Aakhus (Eds.), *Perpetual contact: Mobile communication, private talk, public performance* (pp. 139–169). New York: Cambridge University Press.

Loopt. (2010). Loopt: Discover the world around you. From http://www.loopt.com/.

Løvlie, A. S. (2009). Poetic augmented reality: Place-bound literature in locative media. Paper presented at the Proceedings of the 13th International MindTrek Conference: Everyday Life in the Ubiquitous Era.

Manovich, L. (2002). *The language of new media*. Cambridge, MA: MIT Press.

Manovich, L. (2007). The poetics of augmented space: Learning from Prada. Retrieved May 6, 2007 from http://www.manovich.net/DOCS/Augmented_2005.doc.

Massey, D. (1994). *Place, space, and gender*. Minneapolis: University of Minnesota Press.

Massey, D. (2005). *For space*. London: Sage.

Mateas, M. (2001). A preliminary poetics for interactive drama and games. *Digital Creativity*, 12(3), 140–152.

Meyrowitz, J. (1985). *No sense of place: The impact of electronic media on social behavior*. New York: Oxford University Press.

Meyrowitz, J. (2005). The rise of glocality: New senses of place and identity in the global village. In K. Nyíri (Ed.), *A sense of place: The global and the local in mobile communication* (pp. 21–30). Vienna: Passagen Verlag.

Murray, J. (1997). *Hamlet on the Holodeck: The future of narrative in cyberspace*. Cambridge, MA: The MIT Press.

Nakamura, L. (2007). Race in/for cyberspace: Identity tourism and racial passing on the Internet. In D. Bell, & B. M. Kennedy (Eds.), *The cybercultures reader* (pp. 227–235). New York: Routledge.

Nisi, V., Oakley, I., & Haahr, M. (2008). Location-aware multimedia stories: Turning spaces into places. Paper presented at the Proceedings of ArTech 2008. Retrieved from http://www.whereveriam.org/work/UMa/nisi_oakley_haahr_artech2008_v1.3.pdf.

Oksman, V., & Turtiainen, J. (2004). Mobile communication as a social stage: meanings of mobile communication in everyday life among teenagers in Finland. *New Media & Society*, *6*(3), 319–339.

Palfrey, J., & Gasser, U. (2008). *Born digital*. New York: Basic Books.

Proboscis (Artist). (2002–2004). *Urban tapestries* (Mobile Annotation).

Puro, J. P. (2002). Finland, a mobile culture. In J. Katz, & M. Aakhus (Eds.), *Perpetual contact: Mobile communication, private talk, public performance* (pp. 19–29). Cambridge: Cambridge University Press.

Rheingold, H. (1993). *The virtual community: Homestead on the electronic frontier*. Cambridge, MA: The MIT Press.

Rheingold, H. (2002). *Smart mobs: The next social revolution*. Cambridge, MA: Perseus Publishing.

Scannell, P. (1996). *Radio, television and modern life*. Oxford: Blackwell.

Schivelbusch, W. (1986). *The railway journey: The industrialization of time and space in the 19th century*. Berkeley, CA: University of California Press.

Silverstone, R., & Sujon, Z. (2005). Urban tapestries: Experimental ethnography, technological identities and place. Retrieved from http://www.lse.ac.uk/collections/media@lse/pdf/EWP7.pdf.

Simmel, G. (1950). *The sociology of Georg Simmel* (K. Wolff, Trans.). New York: Free Press.

Solove, D. (2004). *The digital person: Technology and privacy in the information age*. New York: New York University Press.

Sotamaa, O. (2002). All the world's a Botfighters stage: Notes on location-based multiuser gaming. Paper presented at the Proceedings of Computer Games and Digital Cultures Conference.

Standage, T. (2007). *The Victorian internet: The remarkable story of the telegraph and the nineteenth century's on-line pioneers*. New York: Walker & Company.

Sutko, D. M., & de Souza e Silva, A. (2011). Location aware mobile media and urban sociability. *New Media & Society*, *13*(5), 807–823.

Turkle, S. (1984). *The second self: Computers and the human spirit*. New York: Simon and Schuster.

Turkle, S. (1995). *Life on the screen: Identity in the age of the Internet*. New York: Simon and Schuster.

Virilio, P. (1995). *The art of the motor*. Minneapolis: University of Minnesota Press.

Wertheim, M. (1999). *The pearly gates of cyberspace: A history of space from Dante to the Internet*. New York: W. W. Norton & Company.

CONCLUSION: NEAR FUTURES AND THE IMMEDIATE PRESENT

Johanna is finally feeling comfortable in her new city. It has been three weeks since her first day of school, and she has made new friends and begun to enjoy herself. She is also happy about her new Kindle, which her mother had bought her a week ago to celebrate how well she was doing during her first extended time away from home. Johanna loved the Kindle. She already owned an iPad with a Kindle app, and she had been using the app to read books, but after a while the screen hurt her eyes. The Kindle was more like reading a book, and she enjoyed taking it on the bus to read on the way to class. However, she still carries print books in her bag because some of her textbooks aren't available on the Kindle.

On that Monday, Johanna is reading the Kindle on her bus ride when her phone vibrates. She looks down at her phone, annoyed, wishing she had more time to read. However, she checks the phone anyways because she is waiting for an important email from her brother who had told her he was about to propose to his fiancé. She clicks on the email icon and smiles; her email had arrived and it is good news! She begins to type out a response on her iPhone but decides that the small screen is making it too difficult to type a message long enough to reply to the email. She thinks to herself, "I guess I could wait until I get to campus to reply to this email, but I kind of think I'm too excited to do anything else until I reply." Johanna stops for a moment, staring at her phone, and then reaches into her bag and takes out her laptop, which was buried in her bag beneath two heavy textbooks. She opens the laptop and begins drafting a response to the email, knowing that she'll have to wait until she gets to campus to connect to the network and send the email. It normally bothers her that she has to carry so many devices in her bag, but it's times like these that make it all worth it.

Mobile technologies have become an important part of our experience of public spaces. For those who live in urban centers, it is almost impossible to go out on the streets without seeing someone listening to music on an iPod, talking on a mobile phone, or reading a book—maybe even an electronic book. As these technologies become part of the urban landscape, they also challenge established social norms, expectations of privacy, and the very meaning of public spaces. At the most basic level, that has been the central argument of this book. By taking previously private practices out into the streets, mobile technologies allow people to control in new ways their experience of public space, and consequently, blur the boundaries between what is considered private and what is considered public. As we have shown, the act of reading was primarily a private practice because people would generally read inside their houses or in libraries. But when books became portable (lighter and smaller), people could easily carry them to read in parks and on public transportation. The act of reading a book in silence is still mostly private—we often do not share the narrative with strangers around us in public spaces—but the person reading is still a part of the larger interaction site of public space. Readers have been often perceived to be "anti-social" because they refused to interact with other people around them. The same happened with the mobile phone: Telephones were attached to specific places, and telephone conversations generally took place inside the private space of the house or the office. When telephones became portable and conversations were taken into public spaces, many lamented the "privatization" of public spaces by the intrusion of personal voice calls with remote people. Claims of privatization of the public were also applied to personal listening devices, such as the Walkman and the iPod. But even after the book, the Walkman, the iPod, and the mobile phone, public spaces are still public. The way we understand public spaces, however, changed.

Nevertheless, the permeable boundaries between public and private spaces are not solely shaped by mobile technologies. Public spaces have always included private activities. For example, someone might just be sitting alone thinking, or perhaps having a personal meeting with a colleague, or even be strolling through the park on a romantic date, activities that do not directly include other people sharing that same space. Mobile technologies, rather than radically changing how people behave in public, increase the range of personal and private activities they can perform in public spaces. But they also become an intrinsic part of public spaces and do influence how these spaces are configured. They reconfigure the nature of the public and they shape a new experience of public space, but not necessarily by detaching people from it as critics have claimed. Mobile technologies reconfigure public spaces in two major ways: (1) They take practices previously confined to more traditionally private spaces out into the streets of the city, and (2) they give people a feeling of control and familiarity with public spaces typically associated with private spaces. From this perspective, instead of detaching people from spaces or privatizing them, mobile technologies can be viewed as interfaces to public spaces. In this book, we examined how personal and portable technologies (namely books, Walkmans, iPods, mobile phones, and location-aware technologies) help people to manage and control their relationship to public spaces. These mobile technologies have frequently been criticized for working to the detriment of "good" public spaces. But if we think of public spaces as spaces where strangers co-exist (Jacobs, 1961; Sennett, 1977, 1992), then mobile technologies often serve as interfaces people use to control, manage, and filter public sociability.

As we have also shown, people's desire to manage and control their relationships to public spaces has existed for quite a long time, and it is not a by-product of mobile technology use in public. As Simmel (1971) showed, in the early nineteenth century, urbanites developed protection mechanisms to deal with the chaos of urban spaces. He named this type of psychological filter the *blasé* attitude. The use of mobile technologies in public emerged as a different type of filter. Since the early to mid-nineteenth century, books and newspapers became an integral piece in the construction of the shared space of the railway compartment (Schivelbusch, 1986). Faced with a new social situation that forced people to share a space with strangers, passengers turned to the mobile technologies of books and newspapers as a way to negotiate their engagement with both the train space and other people in that space. Over one hundred years later, a new type of personal mobile technology was created that merged music and mobility. The Sony Walkman was released in 1979 and allowed people to impart a personal soundtrack on their experience of public space. They could control, to a degree, their experience of space by replacing the sounds of a space with the music they wanted to hear (Bull, 2000, 2004; du Gay, Hall, Janes, Mackay, & Negus, 1997; Hosokawa, 1984). This form of auditory control also provides a means of thinking through the way in which mobile telephony is situated within the greater ecology of things and people that make up the public. While mobile telephony actually preceded the Walkman by many decades, the widespread adoption of the mobile phone did not begin until the late 1980s and early 1990s (and even later in the United States) (Agar, 2005; Galambos and Abrahamson, 2002). Like the Walkman, the mobile phone allowed people to engage with a different auditory soundscape from what was otherwise present in a shared space. Unlike the Walkman, however, the mobile phone also allowed people to stay connected to other people who were not physically present (Gergen, 2002; Ling, 2004; Ling & Donner, 2008). The ability to engage with information (whether through music or conversation) that is not seemingly part of the physical space led to many criticisms of both the Walkman and the mobile phone (De Gournay, 2002; Gergen, 2002; Puro, 2002). Still today, mobile technology use in public spaces is viewed with skepticism by those who fear that these devices are pushing us apart, instead of forging meaningful connections to co-present people (Gergen, 2010; Turkle, 2011).

Issues of control over public sociability and the borders between private and public can also be applied to location-aware mobile technologies, but location-aware technologies raise a number of new important questions to the study of mobile technologies. For one, they allow for new ways of controlling and filtering information in public spaces (de Souza e Silva & Frith, 2010b; Frith, in press; Gordon & de Souza e Silva, 2011). They do so by enabling people to access location-specific information that is filtered through the interface of the mobile device. People can use location-based services to access information about locations that match predetermined preferences, even going so far as to map their pre-existing social network onto physical space. Earlier mobile technologies also helped users control their experience of public spaces, but by allowing users to access and filter information that is "attached" to a location—rather than bringing outside information to the public space (e.g., the narrative of a novel, songs played through headphones, a conversation with a distant other)—location-aware mobile technologies connect people to their nearby environment.

Paradoxically, however, even as location-aware mobile technologies apparently empower people by enabling them to filter location-based information that is attached to public spaces, they also allow for new ways individuals can be tracked and surveilled (de Souza e Silva & Frith, 2010a). Many of the social concerns raised around location-aware technology use focus on locational privacy. Increasingly, personal location history is being collected and stored in online databases by mobile phone providers, advertising companies, and governments, often without people's knowledge. Frequently, this location information is used to target users with (sometimes unsolicited) location-based advertisements. Location-aware technologies may contribute not only to new forms of corporate and governmental surveillance, but also to new forms of collateral surveillance. Increasingly, parents use chaperone phones to track the whereabouts of their children, and parole officers track parolees via GPS devices (Boesen, Rode, & Mancini, 2010; Shklovski, Vertesi, Troshynski, & Dourish, 2009). Ultimately, these technologies not only help to construct new meanings of locations through the constant and dynamic presence of location-based information, but they also influence people's mobility through urban spaces. For example, a *Foursquare* user may decide to go to the cafe that is five blocks away instead of the bakery at the corner to buy some coffee just because he might add points to his mayorship and get a free drink. On the way, if he sees another friend checked in nearby, he might consider walking a block away from his destination to personally say hello to this friend. Parents also influence their children's patterns of mobility when the children know the parents have access to their location. In order to avoid problems, they will go to places where they will not get into trouble. As location-aware technologies likely become more widely used, the implications for privacy, power, and sociability will be increasingly important to examine.

Writing about mobile technologies, however, raises an immediate question: How do we discuss mobile technologies when there seems to be new mobile technologies released every month? From the development of portable books and newspapers until the release of the Walkman in the late 1970s, not much changed in the personal mobile technology landscape. There were a few attempts to develop new forms of portable technologies, most notably mobile television (Spigel, 2001), but for the most part the mobile technology landscape was relatively static. That is certainly no longer true. Since the release of the Walkman, we have seen an impressive number of new mobile technologies people use to interface with public space, such as iPods, iPads and iPhones, mobile phones, and e-readers. It is impossible to predict with absolute certainty how these developments will play out, and in the next section we do not attempt to do so. We are not "futurologists," and there are many bright people who have made predictions about technological developments that look rather silly in hindsight. Rather than make bold predictions, however, we instead focus on some of the developments likely to affect the personal mobile technology landscape in the near future.

The Proliferation of Mobile Technologies

Too often, people talk about new mobile technologies such as smartphones or tablet computers as if they radically change social and spatial interactions. But they forget that these technologies are embedded into a larger social, political, and economic framework that also shapes the use of these technologies. Furthermore, when addressing changes caused by the use of new technologies, people often forget that past technologies also shaped the way in which people interacted with their spaces and socialized with other people. We can see that most notably in the way people have often ignored the importance of the book as a mobile technology.

However, while it is important to look back and contextualize the development of newer technologies within former social, political, and technological developments, it is also important to recognize new and unprecedented shifts influenced by these new technologies. The last decade has likely seen the development of more new types of mobile technologies than all of human history pre-2000, leading *Wired* to label the decade of the 2000s the "mobile decade" ("The Mobile Decade," 2009). During the 2000s, mobile phones became nearly ubiquitous, reaching saturation point in many nations (Castells, Fernández-Ardèvol, Qiu, Jack, & Sey, 2007); email became fully mobile through the development of the mobile Internet (McQueen, 2010; Ito, Okabe, & Matsuda, 2005); Apple released the iPod, which became a cultural icon and allowed people to carry entire music libraries (Bull, 2007; Kahney, 2005); laptop computers began to outsell desktop computers ("Getting Wired," 2008); e-readers such as the Amazon Kindle became popular; new mobile gaming platforms such as the Nintendo DS were released; and smartphones such as the Apple iPhone turned mobile phones into miniature computers that could run a diverse set of applications (Goggin, 2009, 2011). In 2010, another important mobile technology—the iPad—was released, signaling the moment when tablet computing entered the mainstream. As we are writing this conclusion, in May 2011, the tablet market has become the new mobile tech frontier, with Apple releasing the iPad 2 and major tech companies such as RIM, Samsung, and Motorola developing competing tablets.

So what will the future of mobile technologies look like? Well, it is difficult to say, mostly because of two contrasting trends: the increasing diversity of mobile devices, and a seamless convergence of devices. There are simply more types of mobile technologies than ever before, and this is unlikely to change in the near future. Many of us probably know someone who carries around a smartphone, an iPod, a laptop, and a Kindle. With the development of tablet computers and new mobile gaming systems, there will likely be people who own many different types of mobile technologies. In a way, it seems that what is happening now is the fulfillment of Mark Weiser's early predictions about ubiquitous computing developed in the mid-1990s at the Xerox PARC lab. Mark Weiser and John Seely Brown envisioned that in the future each person would interact with many "computers" in their daily lives (Weiser & Brown, 1996). That would be in contrast to the personal computing era, when the relationship was one person to one computer, and to the time of the mainframes when many people dealt with a single computer. Weiser and Brown's predictions departed so radically from the computational reality in the 1990s, that for years computer scientists, HCI researchers, social scientists, and humanities scholars have been waiting for the ubiquitous computing era, and failed to recognize that ubiquitous computing is already here, but not quite in the same way that Weiser and Brown envisioned. PARC's prototypes, such as the PARCTab, the PARCPad, and the Live Board were designed to be used inside an office space. As Adam Greenfield (2006) notes, this happened because connectivity in closed spaces was easy to obtain. But as soon as wireless positioning systems started to colonize physical spaces, and location could be obtained outdoors (as is the case with Global Positioning Systems (GPS), Wi-Fi, Wi-Max, and Bluetooth) ubiquitous computing moved outdoors. Most of the emerging mobile technologies today include some kind of location awareness. Smartphones include GPS technology, but location awareness is not restricted to GPS. Most mobile phones can be located by triangulation of waves, iPads and Kindles connect to the Internet via 3G and Wi-Fi. And some iPod models include Wi-Fi. And insofar as technologies have a wireless connection, they can be located and are location-aware.

Weiser's idea of Ubicomp was, among other things, based on location-awareness. The PARC Pad and the Tab tracked the position of PARC employees and transmitted it to a central server, in order to provide them with location-based information. Today, we do have tablets and pads, as Weiser and his colleagues envisioned, but mobile phones became much more ubiquitous as mobile technologies. Weiser and Brown (1996) also envisioned ubiquitous computing (or as they named it, "calm technology") disappearing into the spaces of everyday life. But although there is an increasing number of sensors and hidden surveillance cameras spread around urban spaces that are imperceptible, many of the technologies that create the ubiquitous computing ecology today, such as mobile devices and smartphones, are far from being "calm." On the contrary, it seems that mobile technologies are scorned for increasingly disturbing others and being very visible in public spaces. Ubiquitous computing also did not replace personal computers. Desktop computers still exist, and people still consider laptops very personal. The desktop metaphor, despite being almost 30 years old, does not show signs of going away. And although location awareness is an important characteristic of this new computational environment, researchers still have a hard time coordinating indoors and outdoors locative systems.[1] So, what we see today is an ecology of technologies that permeate our environment, that allow us to be more mobile but at the same time increasingly surveilled. Some of the technologies discussed above, such as iPads or portable gaming devices, are not generally framed as "surveillance" technologies. However, at the moment of this writing, people can run all sorts of LBSNs on iPads, and portable gaming devices often offer multiuser connectivity and people can interact with other players nearby. In short, all these technologies now include online connectivity, which supports different forms of top-down and collateral surveillance. These technologies sometimes disappear in the urban landscape, but they also are very visible and present in other situations—as happens, for example, when somebody receives a phone call on a bus. In sum, the current push in the tech world is to make products that are more mobile and portable. That is unlikely to change in the near future.

The second trend in mobile technology development is toward convergence. This trend is apparently in contrast to the undeniable proliferation of new and diverse types of mobile devices. But at the same time that we acquire more devices, we also have devices that combine functions formerly performed by different technologies. Take smartphones for example. Android smartphones and the iPhone now feature voice calling, web browsing, music storage, relatively good cameras, advanced gaming that is often comparable to games found on mobile gaming systems, near-field communication systems that allow them to act as credit cards, and Kindle apps that let people read books on their phones. The same is true of new tablet computers that combine the functionality of multiple devices inside a single device.

The simultaneous diversity and convergence of mobile technology raises many questions. Will the shift toward tablet computing mean that computing may increasingly be based on applications, such as the iPhone apps? Will mobile gaming, and especially location-based mobile gaming, increasingly be a way individuals interface with public space, whether through smartphones or dedicated game devices? As phone cameras continue to improve and tablets come equipped with cameras, will people still carry around digital cameras? As more and more people purchase smartphones, will they continue to use mobile music players such as the iPod touch? And will the reign of the codex book end now that more and more people are adopting e-readers?

Regardless of how mobile interfaces develop in the future, it is important to always take into consideration how they affect sociability and spatiality. Whether people carry more and more devices or just a few devices that can do everything they need, people will still use mobile technologies as interfaces to public spaces. The increasing use of these technologies will also continue to challenge the lines between the private and the public. Take the debates about the future of the book for example. As more people adopt e-readers, there may be a point in the future where we will see fewer and fewer people reading printed books while in public. What's more, e-readers do include new features such as social annotation that may make reading in public more social in a sense. For example, readers are already able to see sections of the book they are reading that were highlighted by other readers. Additionally, in the future, they might be able to see who in their vicinity is reading the same novel, and share comments about the book with them. More interestingly, location-aware reading applications might provide a potentially enhanced experience of a novel if the reader is using the application in the place the novel describes. However, whether people read on Kindles or paperback novels, they are still developing new interactions with the public space they occupy. Also, let's just note that the codex book is a very efficient interface that has been around for more than a thousand years and will not likely disappear in the near future. People will probably still use different forms of books (printed and electronic) for different purposes for some time. For example, a person might have a Kindle to avoid carrying printed books when traveling, but prefer to read printed books at home.

But some things people can do with e-readers as compared to books are undoubtedly different. With e-readers, people can carry entire libraries with them, signaling a shift similar to the one we saw with the move from the Walkman to the iPod. At the same time, moving toward more digital content as compared to print does raises interesting issues about control over content. People actually often have less control over content when it moves from print to digital. Someone who buys a Kindle book on Amazon does not "own" that content in the same way as with a paperback novel (Doctorow, 2009). She cannot resell the book, and she cannot share it with friends. The issue of ownership of content will be important to explore with future developments of mobile technologies; nonetheless, the core desire to control and manage one's experience of space will likely remain present regardless of which new mobile devices are developed.

As more and more people begin to use these new mobile devices, the issues raised in this book will become even more important. New mobile technologies will be created, but questions about the borders between private and public, the nature of public spaces, privacy, and power will still apply to the study of how we interact with these technologies. The issues we have examined throughout this book will still apply whether someone reads an electronic book or a print book, or whether they listen to music on a smartphone or an iPod. There will be technological advancements, but what is almost certain to be true in the future is that we will continue to see the proliferation of mobile devices people use to interface with public spaces. Today an important mobile technology that uses location-awareness and enables new ways to interact with public space is the smartphone.

Smartphones and Location-based Services

Smartphones became a key piece of the mobile technology landscape because they combine the voice calls of mobile phones, the functionalities of a mini-computer, wireless access, and location-awareness. One of the distinguishing characteristics of smartphones is users' ability to download third-party applications. Many location-aware applications, including most location-based services (LBS) and location-based social networks (LBSNs) run on smartphones. The early roots of these services are mostly in the realms of locative media art and location-based mobile games (LBMGs), but to gain widespread use, people had to begin carrying around mobile devices that were capable of running LBS. The release of the iPhone and the creation of the Apple app store was key to the commercial development of LBS. The app store allowed users to download third-party applications to run on their phones, and developers immediately began creating new applications. As of this writing, there are over 300,000 apps in the Apple store and over 150,000 apps in the Android marketplace (Shanklin, 2011). The types of apps range from location-based services that find nearby gas stations to apps that can turn the mobile phone into a wireless mouse or provide immediate food recall information. As more people adopt smartphones, not only the number of apps will likely increase, but there will be also a more diverse set of LBS users. The diffusion of smartphone technology may have important implications for the social and spatial aspects of location-aware applications we discussed in the latter half of this book. Because LBS allow people to digitally annotate the city, as different groups adopt smartphones there will be a larger and more diverse set of annotations. For example, the majority of *Foursquare* users are young and middle to upper-middle class, and that is represented in the types of places that currently have the most annotations. However, as more diverse groups of people use these types of applications, we can expect to see a more representative set of location-based information that includes more types of venues.

Smartphone adoption is likely to increase fairly rapidly in the coming years. With the growth of the iPhone and the Android OS the number of smartphones in the United States has already increased quickly, and a study by ComScore, a prominent Internet marketing research company, shows that there are now 45.5 million smartphones in the country (Block, 2010). The United States is not a special case; it lags behind some European countries in smartphone adoption. Spain, for instance, recently surpassed Italy as the country with the highest percentage of smartphone users (Comscore, 2011). Equally importantly, a new set of high-quality, highly regarded budget smartphones, such as the Android-powered LG Optimus S, have been released in North America and Europe (Figure 7.1). These phones still require data plans, but they are often offered by phone companies for free, compared to higher end phones that retailed at US$249 or US$199 (with a two-year plan). These new developments might contribute to smartphone adoption in developing parts of the world, such as countries in South America, Africa, and Asia. As smartphone adoption increases in the coming years, new budget phones will hopefully allow less-privileged users to purchase these phones.

The expansion of the smartphone market will likely have a major effect on some of the location-based services and applications we discussed throughout this book. Many of the current applications include social elements. For example, the art project *Textopia* we described in Chapter 6 is now available in the Google and Apple app stores and can be run on these phones. People can now annotate space through *Textopia* and similar services, adding to the digital information present in these hybrid spaces and possibly interfacing with the places they move through in new ways. However, at this early stage in their development, applications such as *Textopia* have only limited usefulness. If a user logs in to *Textopia* in most cities, there are often no texts to follow; not enough people use the application to make it worthwhile. For applications such as *Textopia* to work well, they require a critical mass of users to share in the social annotation of locations. With wider smartphone adoption, it is likely that more people will turn to these services as new ways to build meaningful locations and interface with public space.

We can see that beginning to happen with a different type of application we have discussed in this book: LBSN's *Foursquare*, for example, recently surpassed ten million users worldwide, adding an additional one million users in less than

Figure 7.1 The LG Optimus S for the U.S. market on the Sprint Network. Copyright: KWatson1984.

a month (Foursquare, 2011). However, most of these users are still in the United States. There are now enough *Foursquare* users in the United States that in some major cities, such as New York and San Francisco, there are already a wide range of digital annotations and nearby players. As more and more people adopt location-aware mobile devices, the number of LBSN users will likely increase, and these applications will be an important way to interface with the city as a space where strangers co-exist. According to Metcalfe's law, the value of a network increases exponentially with each new node of the network. Each new user in a network makes the entire network more valuable because the chance of connections increases exponentially (Shapiro, 1999). In the early days of LBSNs, there was hardly anyone using the services, meaning the people who did use them did not have many people with whom to interact. As smartphones begin to become more ubiquitous, the number of people who share their location with friends and annotate physical space will increase. With each additional user, these applications become increasingly important in the construction of public places. They will help users to construct new types of locations embedded with dynamic digital information.

The future of smartphones, however, concerns more than just increased adoption statistics. Another important future development of smartphones as mobile technologies will have to do with technological specifications. When third generation (3G) phones were first introduced, they were supposed to revolutionize what people could do with a mobile phone (Goggin, 2006; Wilson, 2006). Industry experts envisioned people using their mobile phones to easily stream television and movies and all sorts of different content. However, the future, as always, did not go as planned. For years after the development of 3G there were not many devices capable of taking full advantage of the mobile Internet (Nielsen & Fjuk, 2010; Wilson, 2006). Even after smartphones began being released, 3G was still too slow to do many of the things that had been promised (Wilson, 2006). The next generation of mobile networks—4G—once again promises to revolutionize mobile technology usage. The 4G connections on the Verizon LTE network in the United States are typically more than fifteen times faster than 3G connections, with download speeds that often exceed 10 mbps (Bonnington, 2011). However, at the time of this writing, most of the technology branded as 4G uses standards such as LTE (3GPP Long Term Evolution), HSPA+, or WiMax (a type of long-range wireless network), which do not fully comply with 4G standards. So, although mobile phone providers market their networks as 4G, 4G is not quite here yet.

Simone Frattasi, Basak Can, Frank Fitzek, and Ramjee Prasad (2005) differentiate between two visions of 4G technology: linear and concurrent. The linear 4G vision sees 4G as merely an extension of 3G technology that will provide devices with higher connection speeds. The problem with this vision is that "it assumes that the network will have a cellular structure, which basically means that it will be built on the fundamental architecture of the preceding generations of mobile technologies" (Frattasi et al., 2005, p. 1). The linear 4G vision is basically what is currently advertised by mobile phone companies, but as we have mentioned above, the future is not limited to cellular systems and thus "4G has not to be exclusively understood as a linear extension of 3G" (p. 1). In contrast, the concurrent 4G vision views 4G not as one specific type of network, but rather an ecology of different types of wireless networks (LTE, Wi-Fi, Wi-Max, Bluetooth, etc.) and devices (mobile tablets, smartphones, RFID tags, cars, refrigerators, PCs) that can seamlessly connect to and exchange information to each other. This vision is strongly associated with the ubiquitous computing paradigm (Greenfield, 2006; Weiser, 1994, 1999) because it focuses on the development of services, potential for sociability, and the heterogeneity of devices and networks that will permeate our environment. Furthermore, it emphasizes that connection among devices will not be restricted to cellular networks, but actually be extended to short-range communication systems such as Bluetooth and other types of wireless protocols (Frattasi et al., 2005; Frattasi, Fathi, Fitzek, Chung, & Prasad, 2004; Frattasi, Fathi, Fitzek, Prasad, & Katz, 2006; Javaid, Rasheed, Meddour, Ahmed, & Prasad, 2008). However, just as the linear 4G vision is not quite here yet, the same is true for the concurrent 4G vision. While a single mobile phone is able to handle connections through an increasing number of networking technologies (LTE, Wi-Fi, RFID, Bluetooth, and others), different devices are still kept separate in actual practice.

Without true network convergence and devices connectivity, opportunities for application development are still limited, because devices are still not fully able to seamlessly exchange information with each other. For example, Javaid et al. (2008) envision systems where people can use their smartphones and other personal mobile devices to interact with intelligent transport systems (ITS), smart-home networking and cooperative healthcare. Although some of these services are available today, they are not fully integrated. Increases in connection speeds will also influence what users can do with personal mobile devices. For example, location-aware augmented reality applications such as *Layar* are currently limited by slow connections, and mobile television services have not really taken off because of problems streaming video over 3G connections. However, it is likely that faster 4G connections will improve many existing applications and will provide individuals with new ways to engage with public spaces. Developers will also likely take advantage of the new technological capabilities of more powerful smartphones, increased networking connectivity, and their connections with other technologies and networks to design applications that further alter the way in which people use these devices to interface with public space.

Public Spaces and Locations

Regardless of the types of applications or specific technologies developed, mobile technologies will enable new forms of relationships to public spaces. Whether through the mapping of social networks, accessing location-specific information, annotating physical space, listening to music, or reading an electronic book, mobile devices enable new ways to engage (and disengage) with shared spaces. While it is important to realize that the way people use technologies to interface with public space is not new, we should also acknowledge the changes that emerge with newer types of mobile technologies.

In the latter half of this book, we specifically focused on location-aware mobile technologies. It is not only the mobility, but also the location-aware aspects of devices that makes them critical interfaces for sociability in public spaces. Different from other mobile technologies that are mostly used individually (such as the iPod) or connect people with remote others (such as the mobile phone), location-aware devices have the potential to connect people with others nearby. On the map on their iPhone, *Foursquare* players can see other players nearby and choose to contact them if they wish. As discussed in Chapter 6, LBSNs might foster connections with unknown people in public spaces because they help people with similar interests find each other. Following the same logic, location-based dating applications, such as *Skout*, *MeetMoi*, and *4Singles* are gaining popularity. Different from online dating sites, they allow users to find matches that are physically close to them. *MeetMoi*, for example, runs constantly on the background of Android phones and alerts users whenever somebody that matches their profile is within a mile of their location.

It is not only urban sociability that will be influenced by location-aware technology use; public spaces themselves will change through the proliferation of networked connections. Throughout the second half of this book, we have seen the increased importance of location when nearly every place can be pinpointed on a map and is embedded with location-based information. *Location* becomes an important term when we refer to our environment and when we socialize with other people. People are worried about locational privacy, they track other people's locations, and they can check into locations. Increasingly, digital information acquires lat/long coordinates and permeates public spaces. People find others depending on their geographical distance to each other, and public places such as restaurants and bars become points on a map. These locations will tend to become more visible, while other places without lat/long coordinates will tend to be less visible—or "findable" on a map. Equally importantly, the information "attached" to locations will also be filtered through the screens of mobile devices, leading to a more personalized experience of public space.

As more people visually interact with their surrounding space through the maps of their mobile devices, we need to think through the consequences of new forms of "screening" the city. Receiving personalized information is not only a consequence of mobile technology use. In his book *The Googlization of Everything*, Siva Vaidhyanathan (2011) showed that people are increasingly receiving personalized search results when using search engines such as Google.

For example, two people performing the exact same search will likely receive different results depending on their IP address and their past browsing history. The increasing personalization of information accessed via the Web can potentially fracture knowledge and lead to further tendencies to seek out sameness when searching for terms. For example, two people who have Google News as their browser homepage will have access to different news depending on the filters they apply, such as their location and preferences for specific topics, but also depending on factors that they might not be aware of, such as the links they click and their social networking connections. With the ability to attach dynamic digital information to locations, we will witness something similar in public spaces. People will be able to personalize public space. But this personalization does not only occur according to users' agency and conscious preferences. Public spaces will be personalized and filtered in ways people will have little control over. For example, because of unclear privacy policies, the types of people in somebody's social networks, or someone's location, each user of these services will have access to a configuration of public space that is unique to them. This leads to experiences of the public that are significantly different among people who use location-aware technologies, and also between people who use these technologies and those who do not. Even some of the most interesting applications we have discussed, such as *Textopia*, could still lead to a fractured type of public space where different people experience different location-based narratives. Personalization has been a key push in the tech world for some time now, and with the adoption of location-aware technologies, the forms of personalization that used to be confined to online spaces have moved into the public spaces of the city (Frith, in press).

Location-aware mobile technologies will contribute to new forms of personalization, but they also will have other influences on the way in which people interface with public spaces. Locations will become increasingly meaningful, as more and more people are able to annotate them and retrieve location-specific information, and therefore actively contribute to the construction of the informational landscape that permeates public spaces. Locations will also be the interface to many social relationships, as people will use location-based applications to find others nearby. As more and more digital information is organized by physical location, the interface people use to access and produce this information will become increasingly important. Public spaces and urban sociability are always in a process of change. But the use of mobile interfaces in public spaces will always be a constant part of urban sociability.

After a year of study abroad, Johanna has been back in her home country for two months. She misses her friends, especially Mark. He said he was going to visit her at some point, but she didn't know when. Little does she know that he is already in town, trying to surprise her. He knows she likes to run at the end of the afternoon. They used to compete with each other by comparing their running paces on the *MapMyRun* app on their iPhones. It is 5 pm on a beautiful sunny day. He gets his iPhone and thinks, "Time to go for a run." He opens the live tracking feature of the *MapMyRun* app and, to his surprise, sees that Johanna is half a kilometer away, heading to the park. "I'll try to catch her," he thinks. "If I run fast enough, I'll catch her. . ."

Note

1   For example, while Wi-Fi and Bluetooth are good locative technologies indoors, their reach is very limited outdoors. They are also not very precise. In contrast, GPS is precise in locating things outdoors, but it works poorly inside the four walls of a building. GPS signal also often fails near tall brick and concrete buildings and during cloudy days.

References

Block, B. (2010). UK leads European countries in smartphone adoption with 70% growth in past 12 months. *Comscore.*

Boesen, J., Rode, J., & Mancini, C. (2010). *The domestic panopticon: Location tracking in families.* Paper presented at the UbiComp '10.

Bull, M. (2000). *Sounding out the city: Personal stereos and the management of everyday life.* Oxford: Berg.

Bull, M. (2004). Thinking about sound, proximity, and distance in Western experience: The case of Odysseus's Walkman. In V. Erlmann (Ed.), *Hearing cultures: Essays on sound, listening, and modernity.* New York: Berg.

Bull, M. (2007). *Sound moves: iPod culture and urban experience.* New York: Routledge.

Castells, M., Fernández-Ardèvol,, M., Qiu, M., Jack, L., & Sey, A. (2007). *Mobile communication and society: A global perspective.* Cambridge, MA: MIT Press.

Comscore. (2011). Smartphone adoption increased across the U.S. and Europe. *Comscore.*

De Gournay, C. (2002). Pretense of intimacy in France. In J. Katz, & M. Aakhus (Eds.), *Perpetual contact: Mobile communication, private talk, public performance* (pp. 193–205). Cambridge: Cambridge University Press.

de Souza e Silva, A., & Frith, J. (2010a). Locational privacy in public spaces: Media discourses on location-aware mobile technologies. *Communication, Culture & Critique, 3*(4), 503–525.

de Souza e Silva, A., & Frith, J. (2010b). Locative social mobile networks: Mapping communication and location in urban spaces. *Mobilities, 5*(4), 485–505.

du Gay, P., Hall, S., Janes, L., Mackay, H., & Negus, K. (1997). *Doing cultural studies: The story of the Sony Walkman*. London: Sage.

Entman, R. M., & Herbst, S. (2001). Reframing public opinion as we have known it. In L. W. Bennett, & R. M. Entman (Eds.), *Mediated politics: Communication in the future of democracy*. New York: Cambridge University Press.

Foursquare. (2011). Wow! The Foursquare community has over 10,000,000 members! From http://blog.foursquare.com/2011/06/20/holysmokes10millionpeople/.

Frattasi, S., Can, B., Fitzek, F. H. P., & Prasad, R. (2005). Cooperative services for 4G. *Proceedings of the 14th IST Mobile & Wireless Communications Summit, Dresden, Germany*.

Frattasi, S., Fathi, H., Fitzek, F., Chung, K., & Prasad, R. (2004). *4G: A user-centric system*. From http://citeseerx.ist.psu.edu/viewdoc/download?doi=10.1.1.59.1748&rep=rep1&type=pdf.

Frattasi, S., Fathi, H., Fitzek, F. H. P., Prasad, R., & Katz, M. D. (2006). Defining 4G technology from the users' perspective. *Network, IEEE, 20*(1), 35–41.

Gergen, K. (2002). The challenge of absent presence. In J. Katz, & M. Aakhus (Eds.), *Perpetual contact: Mobile communication, private talk, public performance* (pp. 227–241). New York: Cambridge University Press.

Gergen, K. (2010). Mobile communication and the new insularity. *Interdisciplinary Journal of Technology, Culture and Education, 5*(1).

Goggin, G. (2006). *Cell phone culture: Mobile technology in everyday life*. London, New York: Routledge.

Goggin, G. (2009). Adapting the mobile phone: The iPhone and its consumption. *Continuum: Journal of Media & Cultural Studies, 23*(2), 231–244.

Goggin, G. (2011). *Global mobile media*. New York: Routledge.

Gordon, E., & de Souza e Silva, A. (2011). *Network locality: How digital networks create a culture of location*. Boston, MA: Blackwell Publishers.

Greenfield, A. (2006). *Everyware: The dawning age of ubiquitous computing*. London: New Riders.

Hosokawa, S. (1984). The Walkman effect. *Popular Music 4*, 165–180.

Ito, M., Okabe, D., & Matsuda, M. (Eds.). (2005). *Personal, portable, pedestrian: Mobile phones in Japanese life*. Cambridge, MA: The MIT Press.

Jacobs, J. (1961). *The death of life of great American cities*. New York: Random House.

Javaid, U., Rasheed, T., Meddour, D. E., Ahmed, T., & Prasad, N. R. (2008). A novel dimension of cooperation in 4G. *Technology and Society Magazine, IEEE, 27*(1), 29–40.

Ling, R. (2004). *The mobile connection: The cell phone's impact on society*. San Francisco: Morgan Kaufman.

Ling, R., & Donner, J. (2008). *Mobile phones and mobile communication*. London: Polity Press.

Nielsen, P., & Fjuk, A. (2010). The reality beyond the hype: Mobile internet is primarily an extension of PC-based internet. *The Information Society, 26*(5), 375–382.

Puro, J. P. (2002). Finland, a mobile culture. In J. Katz, & M. Aakhus (Eds.), *Perpetual contact: Mobile communication, private talk, public performance* (pp. 19–29). Cambridge: Cambridge University Press.

Schivelbusch, W. (1986). *The railway journey: The industrialization of time and space in the 19th century*. Berkeley, CA: University of California Press.

Sennett, R. (1977). *The fall of public man*. New York: Knopf.

Sennett, R. (1992). *The conscience of the eye: The design and social life of cities* New York: Norton.

Shanklin, W. (2011). Number of apps in Android market rapidly gaining on Apple's app store, but Android tablets aren't so lucky. *Android OS, Application News*. From http://www.androidpolice.com/2011/03/14/number-of-apps-in-android-market-rapidly-gaining-on-apples-app-store-but-android-tablets-arent-so-lucky/.

Shklovski, I., Vertesi, J., Troshynski, E., & Dourish, P. (2009). The commodification of location: Dynamics of power in location-based systems. Paper presented at the Proceedings of the 11th International Conference on Ubiquitous Computing.

Simmel, G., & Frisby, D. (2004). *The philosophy of money* (3rd enl. ed.). London, New York: Routledge.

Turkle, S. (2011). *Alone together: Why we expect more from technology and less from each other.* New York: Basic Books.

Vaidhyanathan, S. (2011). *The googlization of everything (and why we should worry).* Berkeley, CA: University of California Press.

Weiser, M. (1994). The world is not a desktop. *Interactions,* (January), 7–8.

Weiser, M. (1999). The computer for the 21st century. *SIGMOBILE Mob. Comput. Commun. Rev.,* *3*(3), 3–11.

Weiser, M., & Brown, J. S. (1996). *The coming age of calm technology.* Xerox PARC.

Wilson, J. (2006). 3G to Web 2.0? Can mobile telephony become an architecture of participation? *Journal of Research into New Media Convergence: The International Technologies,* *12*(2), 229–242.

INDEX